Jan Przewlocki

Simulative Investigation of Post-Oxidation in the Exhaust Manifold of SI Engines

Springer Vieweg

Jan Przewlocki
Institute of Automotive Engineering,
Chair in Automotive Powertrains
University of Stuttgart
Stuttgart, Germany

Zugl.: Dissertation Universität Stuttgart, 2021

D93

ISSN 2567-0042 ISSN 2567-0352 (electronic)
Wissenschaftliche Reihe Fahrzeugtechnik Universität Stuttgart
ISBN 978-3-658-36377-2 ISBN 978-3-658-36378-9 (eBook)
https://doi.org/10.1007/978-3-658-36378-9

Responsible Editor: Stefanie Eggert
This Springer Vieweg imprint is published by the registered company Springer Fachmedien Wiesbaden GmbH part of Springer Nature.
The registered company address is: Abraham-Lincoln-Str. 46, 65189 Wiesbaden, Germany

Wissenschaftliche Reihe Fahrzeugtechnik Universität Stuttgart

Reihe herausgegeben von

Michael Bargende, Stuttgart, Deutschland

Hans-Christian Reuss, Stuttgart, Deutschland

Jochen Wiedemann, Stuttgart, Deutschland

Das Institut für Fahrzeugtechnik Stuttgart (IFS) an der Universität Stuttgart erforscht, entwickelt, appliziert und erprobt, in enger Zusammenarbeit mit der Industrie, Elemente bzw. Technologien aus dem Bereich moderner Fahrzeugkonzepte. Das Institut gliedert sich in die drei Bereiche Kraftfahrwesen, Fahrzeugantriebe und Kraftfahrzeug-Mechatronik. Aufgabe dieser Bereiche ist die Ausarbeitung des Themengebietes im Prüfstandsbetrieb, in Theorie und Simulation. Schwerpunkte des Kraftfahrwesens sind hierbei die Aerodynamik, Akustik (NVH), Fahrdynamik und Fahrermodellierung, Leichtbau, Sicherheit, Kraftübertragung sowie Energie und Thermomanagement – auch in Verbindung mit hybriden und batterieelektrischen Fahrzeugkonzepten. Der Bereich Fahrzeugantriebe widmet sich den Themen Brennverfahrensentwicklung einschließlich Regelungs- und Steuerungskonzeptionen bei zugleich minimierten Emissionen, komplexe Abgasnachbehandlung, Aufladesysteme und -strategien, Hybridsysteme und Betriebsstrategien sowie mechanisch-akustischen Fragestellungen. Themen der Kraftfahrzeug-Mechatronik sind die Antriebsstrangregelung/Hybride, Elektromobilität, Bordnetz und Energiemanagement, Funktions- und Softwareentwicklung sowie Test und Diagnose. Die Erfüllung dieser Aufgaben wird prüfstandsseitig neben vielem anderen unterstützt durch 19 Motorenprüfstände, zwei Rollenprüfstände, einen 1:1-Fahrsimulator, einen Antriebsstrangprüfstand, einen Thermowindkanal sowie einen 1:1-Aeroakustikwindkanal. Die wissenschaftliche Reihe „Fahrzeugtechnik Universität Stuttgart“ präsentiert über die am Institut entstandenen Promotionen die hervorragenden Arbeitsergebnisse der Forschungstätigkeiten am IFS.

Reihe herausgegeben von
Prof. Dr.-Ing. Michael Bargende
Lehrstuhl Fahrzeugantriebe
Institut für Fahrzeugtechnik Stuttgart
Universität Stuttgart
Stuttgart, Deutschland

Prof. Dr.-Ing. Hans-Christian Reuss
Lehrstuhl Kraftfahrzeugmechatronik
Institut für Fahrzeugtechnik Stuttgart
Universität Stuttgart
Stuttgart, Deutschland

Prof. Dr.-Ing. Jochen Wiedemann
Lehrstuhl Kraftfahrwesen
Institut für Fahrzeugtechnik Stuttgart
Universität Stuttgart
Stuttgart, Deutschland

Weitere Bände in der Reihe https://link.springer.com/bookseries/13535

Preface

This work was realized during my tenure as a research associate at the Institute of Internal Combustion Engines and Automotive Engineering (IVK) at the University of Stuttgart under the supervision of Prof. Dr.-Ing. M. Bargende.

My deep gratitude goes to Prof. Dr.-Ing. M. Bargende for his outstanding support, guidance and the numerous motivating discussions.

I want to thank Prof. Dr. Eng. Y. Moriyoshi for his interest and for joining the doctoral committee.

I also want to thank Dr. Michael Grill for promoting my interest on the topic, guiding my work and for his endless support.

In particular, I would like to thank my colleagues. Because of them, I have never lost my motivation and they have helped me to work with joy. Particularly noteworthy are Martin Angerbauer, Markus Maul, Feyyaz Negüs, Rodolfo Tromellini, Viktoria Kelich, Cornelius Wagner, Ismail Mir, Dominik Rether and the coffee machine. Special thanks to Rodolfo Tromellini and Viktoria Kelich for their outstanding commitment and support.

I would also like to thank the working group and all the companies that supported the research tasks within the project "Post-Oxidation" defined and financed by the Research Association for Combustion Engines (FVV) e.V.

I would further like to express my deep sense of gratitude to Christine Burkhardt from EnginOS GmbH for initiating and guiding the research project and for her continuous advice.

Finally yet importantly, I am very grateful to my family, my friends and my girlfriend Annika for their belief and support. They never doubted any of my decisions and helped me to clear my mind in stressful times. Special thanks to all participants in the proofreading competition and congratulations to the winner Annika Guderian.

Stuttgart

Jan Przewlocki

Contents

Figures

Tables

Abbreviations

CFD	Computational Fluid Dynamics
CFL	Courant–Friedrichs–Lewy condition
DISI	Direct Injection Spark Ignition
EGR	Exhaust Gas Recirculation
FKFS	Forschungsinstitut für Kraftfahrwesen und Fahrzeugmotoren Stuttgart
FVV	Forschungsvereinigung Verbrennungskraftmaschinen e. V.
HC	Hydrocarbons
IVK	Institut für Verbrennungsmotoren und Kraftfahrwesen
RDE	Real Driving Emissions
TDCF	Top Dead Center Firing
TWC	Three-Way Catalyst

Symbols

	Latin Letters	
A, B, C, D	Reactants	-
$[A], [B], [C], [D]$	Reactants Concentration	mol/m^3
a_1-a_7	NASA Polynomial Coefficients	-
A_{Fl}	Flame Surface	m^2
b	Temperature Exponent	-
c_p	Specific Heat Capacity, Constant Pressure	J/(kg K)
E_A	Activation Energy	J
H^0	Molar Specific Enthalpy at Standard Conditions	J/mol
K_C	Equilibrium Constant — Concentration based	-
k_f	Forward Reaction Rate	-
K_P	Equilibrium Constant — Pressure based	-
k_r	Reverse Reaction Rate	-
L_{st}	Stoichiometric Air-Fuel Ratio	-
M	Inert Reaction Partner	-
M	Mixing Level	-
m_{air}	Air Mass	kg
m_b	Mass of the Burnt Zone	kg
$m_{(CO+H_2)_{mixed}}$	Combined and Accumulated Mixed CO and H_2 mass	kg
$m_{(CO+H_2)_{unmixed}}$	Combined and Accumulated Unmixed CO and H_2 mass	kg
$m_{(CO+H_2)_{total}}$	Combined and Accumulated CO and H_2 mass	kg
m_e	Mass Entering the Flame Zone	kg
m_F	Mass of the Flame Zone	kg
m_{fuel}	Fuel Mass	kg
m_{ub}	Mass of the Unburnt Zone	kg
n	Amount of Substance	mol
p	Pressure	bar
ppm	Parts per Million	-

R	Specific Gas Constant	J/(kg K)
R_u	Gas Constant	J/(mol K)
rpm	Revolution per Minute	1/min
S^0	Molar Specific Entropy at Standard Conditions	J/(mol K)
T	Temperature	K
t	Time	s
$T_{Ex.Gas}$	Temperature of the Exhaust Gas	K
T_{Wall}	Temperature of the Exhaust Manifold Walls	T
u_e	Flame Front Speed	m/s
u_{fluid}	Fluid Speed	m/s
u_L	Laminar Flame Speed	m/s
u_{sound}	Speed of Sound	m/s
u_{turb}	Isotropic Turbulent Flame Speed	m/s
V	Volume	m^3
v	Evaluation Function	-
Z_{H_2}	Ratio of remaining H_2 after 5 ms	-
Z_{CO}	Ratio of remaining CO after 5 ms	-

Greek Letters

N	Oxidation-Limit Number	-
ι	Ratio of Available and Required Oxidizer	-
λ	Air – Fuel Ratio	-
$\lambda_{Combustion}$, $\lambda_{Comb.}$	Air – Fuel Ratio of the In-Cylinder Combustion	-
$\lambda_{Post-Oxidation}$, $\lambda_{Post-Ox.}$	Air – Fuel Ration of the Post-Oxidation	-
ν_i	Stoichiometric Coefficient of the i-th Species	-
ρ_{ub}	Density of the Unburnt Zone	kg/m^3
τ_L	Characteristic Burn Time	s
χ_i	Mole Fraction of the i-th Species	mol/mol

Indices

b	Burnt
Combustion, Comb.	Related to the In-cylinder Combustion

e	Entering
Ex.Gas	Exhaust Gas
f	Forward
Fl	Flame
Post-Oxidation, Post-Ox.	Related to the Post-Oxidation
r	Reverse
st	Stoichiometric
turb	Turbulent
ub	Unburnt

Abstract

With the introduction of emissions measurements in real driving conditions, the need for improved transient engine behavior while minimizing emissions is increasing. Turbocharged gasoline engines can deliver the same rated power as conventional naturally aspirated engines but with reduced displacement. The exhaust gas enthalpy is used to drive a turbine, which in turn drives a compressor. This then increases the available fresh air mass flow. The result is a higher power density. At operating points with low exhaust gas mass flow or low exhaust gas enthalpy, the available torque of the engine is limited. This is particularly the case at operating points with low engine speed. The reason for this is the low efficiency of the turbine at low mass flow. In addition to the available exhaust gas enthalpy, the turbine power that can be extracted from it also decreases at low engine speed. Due to the direct coupling of turbine and compressor, the compressor power also decreases accordingly. A low compressor output results in a low fresh air mass flow. The result is again lower total engine mass flow, followed by again reduced turbine efficiency. This chain of causality cannot be broken by the engine without additional impulse, and thus the maximum engine torque is limited at low engine speeds.

One measure to improve transient engine behavior in the low speed range is the use of scavenging air. In this case, fresh air is supplied directly from the intake side to the exhaust side through a valve overlap, without this scavenging air participating in the combustion. This leads to an increase in the exhaust gas mass flow and thus also to an increase in turbine efficiency. The direct coupling of turbine and compressor thus also increases the fresh air mass flow. The causality chain described above can be broken, and a self-reinforcing effect sets in that raises the maximum torque.

Due to the way the three-way catalyst (TWC) works, the efficiency of exhaust gas aftertreatment in gasoline engines is heavily dependent on the composition of the exhaust gas. The catalyst achieves the highest conversion rate with a stoichiometric composition of fresh air and supplied fuel. If the composition deviates from this, the conversion rate drops sharply. By using scavenging air to increase the mass flow while the combustion in the cylinders remains unchanged, the composition in the exhaust manifold deviates from the desired stoichiometric ratios. To ensure the high conversion rate of the catalyst, combustion in the cylinders takes place under rich conditions. As a result, the combustion products additionally contain incompletely converted fuel in the form of carbon monoxide (CO) and hydrogen (H_2). Interaction with the unburned scavenging air can lead to an oxidation process in the exhaust manifold and thus to heat release. This so-called post-oxidation process leads to an increase in the exhaust enthalpy, which results in a further increase in turbine power.

Previous approaches of post-oxidation investigation or post-oxidation modelling have mostly been experimentally oriented or carried out by means of 3D-CFD simulation, which entails a high computational time. The use of scavenging air is subject to an adaptation of the engine operating strategy (e.g. in terms of valve timing). The design of an operating strategy or the modification of an already existing one requires a simulation model which is characterized by short computation times. The 1D simulation domain is particularly well suited for this purpose. Thus, many variants can be investigated and the necessary parameter adjustments can be optimized. However, for the possible subsequent post-oxidation effects within the exhaust manifold, there is no reliable model developed on the basis of detailed investigations of the phenomenon. The development of a 1D post-oxidation model within the scope of this work now creates the possibility of being able to represent the use of scavenging air more realistically than before in the 1D simulation environment, without having to rely on time-consuming 3D-CFD simulations or expensive test bench measurements.

This work addresses the development of a 1D post-oxidation model based on validated 3D-CFD simulation results. The 3D-CFD simulations include the exhaust manifold of a four-cylinder gasoline engine and are coupled with a reaction mechanism that allows the reaction kinetic representation of a $CO/H_2/O_2$ system. These raw 3D-CFD data are evaluated and based on these findings, a 1D post-oxidation model is designed, capable of representing the heat release and species transformation in the exhaust manifold.

The post-oxidation processes in the exhaust manifold are mainly characterized by mixing effects and the speed of the oxidation reactions of CO and H_2. In order to be able to evaluate the respective influence of mixing and reaction kinetics, an indicator is being developed. This oxidation-limit number enables the evaluation of the respective influence of mixing and oxidation rate on the limits of post-oxidation within the exhaust manifold. This newly developed indicator makes it clear that the respective limiting influence of lack of mixing and slow reaction rate depends on different boundary conditions.

The investigations show that both effects need to be represented by a post-oxidation model. Since both processes have a simultaneous influence on post-oxidation, the 1D post-oxidation model is divided into two sub-models for mixing and oxidation.

The mixing model refers to 3D-CFD simulation results, which do not include a reaction mechanism. This means that the mixing within the exhaust manifold is not overlaid by oxidation effects. The characteristic mixing behavior of the exhaust manifold under investigation can then be mapped using a newly developed mixing indicator via the discretization of the 1D flow model.

The oxidation of CO and H_2 mixed with fresh air is predominantly controlled by the temperature in the exhaust manifold. Due to the resulting heat release, this process is subject to a highly self-reinforcing effect. The oxidation model is based on the development of a surrogate mechanism similar to the reaction mechanism used in the 3D-CFD simulation. However, since the 1D post-oxidation model is intended to combine precision and speed, this surrogate mechanism must be significantly reduced while compensating for the effect of lack of radial temperature resolution in the exhaust manifold. The surrogate mechanism for CO and H_2 oxidation that emerges from this work involves two reaction equations and a smoothing effect. This smoothing effect modifies the abruptly increasing reactivity in a way that, in the case of the 3D-CFD simulation, the radial temperature distribution is responsible for (which cannot be represented within the 1D model class). The oxidation model shows a very good approximation of the reaction mechanism results over a wide range of temperatures and compositions and does not require a tuning parameter.

The resulting 1D post-oxidation model shows very good agreement with the available 3D-CFD simulation results in terms of total heat release and also the respective spatial distribution within the exhaust manifold. The model responds to a sensitivity analysis with a reasonable and comprehensible behavior. The model tuning is limited here to the adaptation of the mixing model to the characteristic mixing of the exhaust manifold used.

Kurzfassung

Mit der Einführung von Emissionsmessungen im realen Fahrbetrieb steigt der Bedarf an einem verbesserten transienten Motorverhalten bei gleichzeitiger Minimierung der Emissionen. Turboaufgeladene Ottomotoren können bei verringertem Hubraum gleiche Nennleistungen abrufen wie herkömmliche Saugmotoren. Hierbei wird die Abgasenthalpie genutzt, um eine Turbine zu betreiben die wiederum einen Verdichter antreibt. Dieser erhöht dann den verfügbaren Frischluftmassenstrom. Das Ergebnis ist eine höhere Leistungsdichte. In Betriebspunkten mit geringem Abgasmassenstrom bzw. einer geringen Abgasenthalpie ist das verfügbare Drehmoment des Motors begrenzt. Dies ist vor allem in Betriebspunkten mit geringer Motordrehzahl der Fall. Die Ursache hierfür ist die geringe Effizienz der Turbine bei geringem Massenstrom. Neben der verfügbaren Abgasenthalpie sinkt bei geringer Motordrehzahl auch die daraus gewinnbare Turbinenleistung. Durch die direkte Kopplung von Turbine und Verdichter sinkt auch die Verdichterleistung entsprechend ab. Eine geringe Verdichterleistung führt zu einem geringen Frischluftmassenstrom. Die Folge ist ein erneut geringerer Gesamtmassenstrom des Motors, gefolgt von einem erneut verringerten Turbinenwirkungsgrad. Diese Kausalitätskette kann vom Motor nicht ohne zusätzlichen Impuls durchbrochen werden und damit ist das maximale Motordrehmoment bei geringen Motordrehzahlen begrenzt.

Eine Maßnahme das transiente Motorverhalten im Bereich geringer Drehzahlen zu verbessern ist der Einsatz von Spülluft. Hierbei wird durch einen Ventilüberschnitt Frischluft direkt von der Ansaugseite der Abgasseite zugeführt, ohne dass diese an der Verbrennung teilnimmt. Das führt zu einer Erhöhung des Abgasmassenstroms und damit zusätzlich zu einer Steigerung des Turbinenwirkungsgrads. Durch die direkte Kopplung von Turbine und Verdichter wird damit auch der Frischluftmassenstrom erhöht. Die beschriebene Kausalitätskette kann durchbrochen werden und es setzt ein sich selbst verstärkender Effekt ein, der das maximale Drehmoment anhebt.

Die Abgasnachbehandlung von Ottomotoren ist aufgrund der Wirkungsweise des Dreiwege-Katalysators (Three-Way Catalyst, TWC) in ihrer Effizienz stark von der Zusammensetzung des Abgases abhängig. Die höchste Umsetzungsrate erreicht der Katalysator bei einer stöchiometrischen Zusammensetzung von Frischluft und zugeführtem Kraftstoff. Bei einer Abweichung von dieser sinkt die Umsetzungsrate stark ab. Durch den Einsatz von Spülluft zur Erhöhung des Massenstroms bei einer unveränderten Verbrennung in den Zylindern weicht die Zusammensetzung im Abgaskrümmer von den gewünschten stöchiometrischen Verhältnissen ab. Um die hohe Umsetzungsrate des Katalysators dennoch gewährleisten zu können, erfolgt die Verbrennung in den Zylindern unter fetten Bedingungen. Dies hat zur Folge, dass die Verbrennungsprodukte zusätzlich nicht komplett umgesetzten Kraftstoff in Form von Kohlenstoffmonoxid (CO) und Wasserstoff (H_2) beinhalten. Bei einer Interaktion mit der unverbrannten Spülluft kann es zu einem Oxidationsprozess im Abgaskrümmer und damit zu einer Wärmefreisetzung kommen. Dieser sogenannte Nachoxidationsprozess führt zu einer Erhöhung der Abgasenthalpie, was eine weitere Steigerung der Turbinenleistung zur Folge hat.

Die bisherigen Ansätze, die Nachoxidation zu untersuchen bzw. zu modellieren, sind meist experimentell ausgerichtet oder erfolgten mittels 3D-CFD Simulation, welche eine hohe Rechendauer mit sich bringt. Der Einsatz von Spülluft unterliegt einer Anpassung der motorischen Betriebsstrategie (z.B. in Bezug auf die Ventilsteuerzeiten). Die Auslegung einer Betriebsstrategie oder die Modifikation einer bereits bestehenden bedarf eines Simulationsmodells, welches sich durch kurze Rechenzeiten auszeichnet. Die 1D hierfür besonders gut geeignet. Dadurch können viele Varianten untersucht werden und die notwendigen Parameteranpassungen optimiert werden. Für die eventuell nachfolgenden Nachoxidationseffekte innerhalb des Abgaskrümmers gibt es jedoch kein verlässliches Modell, das auf Basis detaillierter Untersuchungen des Phänomens entwickelt wurde. Die Entwicklung eines 1D Nachoxidationsmodells im Rahmen dieser Arbeit schafft nun die Möglichkeit, den Einsatz von Spülluft realitätstreuer als bislang in der 1D Simulationsumgebung darstellen zu können, ohne auf zeitintensive 3D-CFD Simulationen oder teure Prüfstandmessungen angewiesen zu sein.

Diese Arbeit befasst sich mit der Entwicklung eines 1D Nachoxidationsmodells, basierend auf validierten 3D-CFD Simulationsergebnissen. Die 3D-CFD Simulationen umfassen den Abgaskrümmer eines Vierzylinder-Benzinmotors und sind mit einem Reaktionsmechanismus gekoppelt, welcher die reaktionskinetische Darstellung eines $CO/H_2/O_2$-Systems ermöglicht. Die Randbedingungen der 3D-CFD Simulation des Abgaskrümmers stammen von einer 3D-CFD Vollmotorsimulation. Die 3D-CFD Simulationsergebnisse des Abgaskrümmers werden ausgewertet und auf Basis dieser Erkenntnisse wird ein 1D Nachoxidationsmodell entworfen, welches fähig ist, die Wärmefreisetzung und die Speziestransformation im Abgaskrümmer abzubilden.

Um die Vorgänge der Nachoxidation im Detail analysieren zu können, wird in dieser Arbeit auf Reaktionskinetik-Simulation zurückgegriffen. Dieser 0D-Simulationsansatz ermöglicht einen Einblick in die chemischen Vorgänge während einer Oxidation von CO und H_2 unter Berücksichtigung von Zeitskalen. Reaktionskinetische Simulationen arbeiten mit Reaktionsmechanismen, welche den Oxidationsprozess basierend auf dessen Elementarreaktionen beschreiben können. Diese Untersuchungen haben gezeigt, dass die Oxidationsgeschwindigkeit von CO neben der Temperatur von der Konzentration des Radikals OH abhängt. Sobald eine gewisse Konzentration von OH erreicht ist (die notwendige Konzentration hängt unter anderem von der Temperatur und dem Druck ab), beginnt der Oxidationsprozess von CO. Die Zeit, welche notwendig ist die notwendige Konzentration von OH zu bilden, ist die chemische Zündverzugszeit.

Sobald der Oxidationsprozess startet, erhöht sich die Temperatur des Systems, was wiederum zu einer zusätzlichen Erhöhung der Radikalbildungsrate führt. Es entsteht eine Kettenreaktion.

Diese Arbeit zeigt zusätzlich, dass ein erhöhter Druck die Zündverzugszeit der Nachoxidation durch die Bildung von HO_2 Moleküle erhöhen kann. Zusätzlich wird der Einfluss von NO auf das $CO/H_2/O_2$-System untersucht.

Die Auswertungen der 3D-CFD Rohdaten zeigen, dass die Nachoxidation innerhalb des Abgaskrümmers ein sich selbst verstärkender Effekt ist. Sobald eine, zum Start des Prozesses notwendige, Grenztemperatur überschritten ist, erhöht sich die Temperatur aufgrund der Reaktionen zusätzlich. Die Folge ist eine noch ausgeprägtere Nachoxidation. Eine Parameterstudie mit veränderter Abgastemperatur zeigt, dass eine Absenkung der Abgastemperatur die Ausbildung des selbst verstärkenden Prozesses verzögert. Bei gleicher Verweilzeit innerhalb des Abgaskrümmers führt das zu einer Reduzierung der freiwerdenden Reaktionsenthalpie. Bei ausreichend langer Verweildauer kann in beiden Fällen eine vollständige Nachoxidation eintreten. Wird die Temperatur des Abgases jedoch stark verringert, kann der Fall eintreten, dass ein Nachoxidationsprozess startet aber durch die fortlaufende Vermischung mit kalter unverbrannter Luft wieder zum Erliegen gebracht wird. In diesem Fall kann die freiwerdende Reaktionsenthalpie die Temperaturabsenkung aufgrund der Vermischung nicht ausgleichen.

Die Nachoxidationsprozesse im Abgaskrümmer sind hauptsächlich von Vermischungseffekten und der Geschwindigkeit der Oxidationsreaktionen von CO und H_2 geprägt. Um den jeweiligen Einfluss von Vermischung und Reaktionskinetik bewerten zu können, wird eine Kennzahl entwickelt. Diese Oxidationslimit-Kennzahl ermöglicht die Auswertung des jeweiligen Einflusses von Vermischung und Oxidationsgeschwindigkeit auf die Limitierungen der Nachoxidation innerhalb des Abgaskrümmers. Durch diese neu entwickelte Kennzahl wird deutlich, dass der jeweils limitierend wirkende Einfluss von fehlender Vermischung und langsamer Reaktionsgeschwindigkeit von verschiedenen Randbedingungen abhängt. Zusätzlich kann beobachtet werden, dass sich diese Limitierung bei unterschiedlichen Positionen innerhalb des Abgaskrümmers unterschiedlich auswirkt. Am Ende eines Abschnittes mit verhältnismäßig geringer Vermischung wird die Nachoxidation stärker durch fehlende Vermischung beeinflusst als durch langsame Reaktionen. Die Oxidationslimit-Kennzahl kann also dazu dienen, an verschiedenen Stellen das Abgaskrümmers die lokal limitierende Größe zu ermitteln um daraufhin mit Gegenmaßnahmen die Nachoxidation zu beeinflussen.

Die Untersuchungen zeigen, dass beide Effekte, Vermischung und Oxidation durch ein Nachoxidationsmodell abzubilden sind. Da beide Prozesse gleichzeitig Einfluss auf die Nachoxidation nehmen, wird das 1D Nachoxidationsmodell in zwei Untermodelle für Vermischung und Oxidation aufgeteilt.

Das Vermischungsmodell bezieht sich hierbei auf 3D-CFD Simulationsergebnisse, welche keinen Reaktionsmechanismus beinhalten. Dadurch wird die Vermischung innerhalb des Abgaskrümmers nicht durch Oxidationseffekte überlagert. Die Untersuchungen zeigen, dass die Vermischung im Abgaskrümmer einer durch die Geometrie bedingten Charakteristik folgt. Vor allem in Bereichen mit veränderlichem Strömungsquerschnitt oder in Bereichen aufeinandertreffender Fluten ist die Vermischung stark erhöht. In Rohrabschnitten mit relativ konstantem Querschnitt ist die Vermischung von Abgas und Spülluft begrenzt.

Diese hier beschriebene Vermischungsauswertung basiert auf dem Vermischungsniveau M. Diese Kennzahl dient dazu, sowohl in der 3D-CFD, als auch in der 1D Simulationsumgebung die Vermischung von Abgas mit Spülluft beschreiben zu können. Um sowohl für die 3D-CFD als auch für die 1D Simulationsumgebung anwendbar zu sein, wird die Vermischung flächengemittelt ausgewertet. Diese Flächenmittelung stellt zwar eine starke Vereinfachung der 3D-CFD Ergebnisse dar, schafft aber eine Möglichkeit der Adaption innerhalb der 1D Simulationsumgebung.

Das charakteristische Vermischungsverhalten des untersuchten Abgaskrümmers kann dann mithilfe eines neu entwickelten Vermischungsindex über die Diskretisierung des 1D Strömungsmodells abgebildet werden. Je gröber die Diskretisierung einer 1D Strömungskomponente, desto größer ist der Bereich der Mittelung von sich darin befindlicher chemischer Spezies. Eine, für die 1D Simulation übliche, relativ grobe Diskretisierung führt zu einer überschätzten Vermischung von Abgas und Spülluft innerhalb des Abgaskrümmers. Eine feinere Diskretisierung ermöglicht die Darstellung der bereits beschriebenen limitierten Vermischung in Rohrabschnitten mit relativ konstanten Querschnitt. Die starke Vermischung innerhalb von Rohrverbindungsbereichen und Bereichen sich stark aufweitender Querschnitte kann unter Zuhilfenahme von Strömungskomponenten erzielt werden, die ein relativ großes Volumen mit nur einem Diskretisierungsgebiet abdecken. So können Gebiete mit starker Wirbelbildung in der 1D Simulationsumgebung imitiert werden.

Die Oxidation des mit Frischluft vermischten CO und H_2 ist überwiegend durch die Temperatur im Abgaskrümmer gesteuert. Durch die resultierende Wärmefreisetzung unterliegt dieser Prozess einem hochgradig selbstverstärkenden Effekt. Das Oxidationsmodell basiert auf der Entwicklung eines Ersatzmechanismus, ähnlich dem in der 3D-CFD Simulation eingesetzten Reaktionsmechanismus. Da das 1D Nachoxidationsmodell Präzision und Schnelligkeit verbinden soll, muss dieser Ersatzmechanismus jedoch deutlich reduzierter sein und gleichzeitig den Effekt fehlender radialer Temperaturauflösung im Abgaskrümmer ausgleichen. Der aus dieser Arbeit hervorgehende Ersatzmechanismus für die CO und H_2 Oxidation beinhaltet zwei Reaktionsgleichungen und einen Glättungseffekt. Dieser Glättungseffekt modifiziert die sprunghaft ansteigende Reaktivität in einer Weise, wie es im Falle der 3D-CFD Simulation die radiale Temperaturverteilung zu verantworten hat (welche innerhalb der 1D Modellklasse nicht abgebildet werden kann).

Das Oxidationsmodell zeigt über einen breiten Temperatur- und Zusammensetzungsbereich eine sehr gute Annäherung an die Ergebnisse des Reaktionsmechanismus und benötigt keinen Abstimmungsparameter.

Das resultierende 1D Nachoxidationsmodell zeigt eine sehr gute Übereinstimmung mit den zur Verfügung stehenden 3D-CFD Simulationsergebnissen in Bezug auf die Gesamtwärmefreisetzung und auch auf die jeweilige räumliche Verteilung innerhalb des Abgaskrümmers. Auf eine Sensitivitätsanalyse reagiert das Modell mit einem sinnvollen und nachvollziehbaren Verhalten. Die Modellabstimmung beschränkt sich hierbei auf die Anpassung des Vermischungsmodells auf die charakteristische Vermischung des verwendeten Abgaskrümmers.

Es ist jedoch notwendig, die Ergebnisse des Modells unter Berücksichtigung der Modellklasse zu beurteilen. Ein 1D Nachoxidationsmodell kann per Definition nicht in der Lage sein, die dreidimensionalen Effekte der Vermischung nachzubilden. Die Nachbildung der Vermischung imitiert lediglich das Ergebnis der Vermischung durch Wirbel und Strömungsablösungen und nicht die Ursache an sich. Zusätzlich dazu, ist die in dieser Arbeit entwickelte Methode zur Auswertung der Vermischung stark vereinfacht, um auch in der 1D Modellklasse anwendbar zu sein.

Nichtsdestotrotz, das 1D Nachoxidationsmodell ermöglicht eine Darstellung des Nachoxidationeffekts im Abgaskrümmer. Dabei ist es ca. 63 mal schneller als der für diese Arbeit gewählte 3D-CFD Ansatz. Das ermöglicht den Einsatz in einem kompletten Motorkennfeld und damit die Entwicklung einer Motorbetriebsstrategie unter Zuhilfenahme von Scavenging und Nachoxidation.

1 Introduction

The introduction of real driving emission measurements increases the need for improved transient engine behavior while keeping the emissions minimum. The downsizing of direct injection spark ignition (DISI) engines using a turbocharging system leads to delayed torque build-up at transient operation (also known as turbo lag). The turbo lag origins in the lack of mass flow through the turbine or too low exhaust gas enthalpy in front of the turbine to operate the turbine in areas with high efficiency.

A possible way to overcome this lack of mass flow is using scavenging in engine operating points with limited mass flow rate, for example at low engine speed. During scavenging operation, the intake valve opens without the exhaust valve already being closed. This enables a mass flow of fresh air from the intake manifold directly into the exhaust manifold without participating in any in-cylinder combustion. By doing so, the mass flow at the turbine increases and the operating point shifts to an area with higher efficiency. Since this entails an increased boost pressure, the engine is able to increase its cylinders charge and the turbines efficiency can be increased even more. Using scavenging air can rock up the engine load at a constant low engine speed significantly.

The aftertreatment device used for this kind of engines is the three-way catalyst (TWC). This device allows the conversion of undesirable emission species like hydrocarbons (HC), CO and NO_X into H_2O, N_2 and CO_2. This is made by three basic reactions in which NO and CO are converted into N_2 and CO_2 and also CO and HC emissions are reduced by oxidizing. This emission conversion is very effective as long as the conditions are very close to stoichiometry. Lack of O_2 leads to a reduced CO and HC conversion rate while an excess of O_2 leads to an increased CO conversion. As a result, the NO lacks of reaction partner. The outcome is a high NO concentration in the exhaust gas.

To avoid this NO emissions while using the scavenging air to increase the mass flow rate at the turbine, an increased amount of CO is necessary. This is achieved using a rich combustion inside the cylinders. The overall engine can

J. Przewlocki, *Simulative Investigation of Post-Oxidation in the Exhaust Manifold of SI Engines*, Wissenschaftliche Reihe Fahrzeugtechnik Universität Stuttgart,
https://doi.org/10.1007/978-3-658-36378-9_1

then be operated in stoichiometric condition and the TWC conversion rate can be ensured.

The interaction of these incomplete burnt exhaust gas with the scavenging air inside the exhaust manifold can cause an oxidation process. Since this oxidation process occurs after the actual combustion inside the cylinder it is called post-oxidation. This post-oxidation process allows the transformation of chemically bound energy into an increasing enthalpy of the exhaust gas. This increased enthalpy leads to a higher turbine efficiency. However, this methodology need to take the circumstances inside the exhaust manifold into account. The rich incomplete burnt combustion products might approach the turbine without interacting with the scavenging air, resulting in a not increased enthalpy.

Since this is a promising approach in improvement of the engines transient behavior, it's necessary to develop an engine operating strategy including scavenging and post-oxidation. The state of the art simulation methodology of engine operating strategy development is the 0D/1D simulation domain. The interaction of exhaust gas and scavenging air is supposed to be a complex three dimensional phenomenon and therefore the 0D/1D approach is at the current state unable to cover these effects. The convoluted interaction of chemical species during the oxidation processes requires a high calculation effort and this counteracts the main advantage of the 0D/1D simulation approach, its low simulation time.

A simulation approach covering the most important aspects of the post-oxidation inside the exhaust manifold is necessary to being able to rate the post-oxidation process within a 0D/1D simulation.

In the following, detailed 3D-CFD simulation results including reaction kinetics are evaluated. The most important influencing effects are investigated and these findings are used to develop a 0D/1D post-oxidation model approach.

2 Fundamentals and State of the Art

Downsizing is a continuing trend in engine development. With an equal engine speed, the engine operating point shifts toward higher loads, leading to a de-throttling effect and higher engine efficiency. Using a turbocharger system to increase the pressure in front of the cylinders allows an increased cylinder filling and a similar engine torque with less involved cylinders. A compressor ensures the increase in boost pressure. This compressor again is connected to a turbine via a shaft. The turbine is operated by using the mass flow of the exhaust manifold and its high enthalpy. A turbocharged engine with a similar displacement as a naturally aspired engine can be operated with a highly increased rated power.

The high power output can be maintained as long as the compressor can maintain the increased boost pressure in front of the cylinders. Since the compressor is dependent on the turbine speed, the boost pressure is directly connected to the mass flow rate and the exhaust gas enthalpy. Engine operating points characterized by a rather low exhaust gas mass flow, generate a limited turbine speed. This inhibits the increase in boost pressure and therefore, an increasing exhaust gas mass flow. Typical operating points with low exhaust gas mass flow are low engine speed operating points. This results in a major disadvantage of the downsized turbocharged spark-ignition engines. The rated power might be similar to a naturally aspired engine with higher displacement, but as soon as an engine specific engine speed is undercut, the exhaust gas mass flow rate cannot increase the turbine speed due to its low efficiency with low mass flow rates. The power transmitted to the compressor is then also rather low, resulting in a low mass flow. The low mass flow leads again to a decrease in turbine efficiency. This results in a decreasing maximum engine torque with decreasing engine speed. The engine operating point with the lowest engine speed and still achieved maximum engine torque is called low-end torque.

To improve the transient engine behavior, the low-end torque needs to be shifted towards a lower engine speed. This low-end torque shift can be achieved by various methods.

J. Przewlocki, *Simulative Investigation of Post-Oxidation in the Exhaust Manifold of SI Engines*, Wissenschaftliche Reihe Fahrzeugtechnik Universität Stuttgart,
https://doi.org/10.1007/978-3-658-36378-9_2

Since the origin of the problem is the turbine's low efficiency for low mass flow rates, a possible solution is the shift of the turbine's efficiency maximum to lower mass flow rates. By decreasing the turbine size, a lower exhaust gas mass flow or a mass flow with a lower enthalpy is able to maintain operating point with higher efficiency. Then again, a smaller turbine cannot handle the same maximum amount of incoming mass flow or high enthalpy of the mass flow. As a result, parts of the mass flow need to be bypassed using a wastegate to avoid a thermal choking. The bypass mass flow limits the maximum turbine power and therefore, the maximum compressor power. This lowers the rated power of the engine.

Another possibility of shifting the efficiency maximum, without limiting the rated power significantly, is by using a variable geometry turbocharger (VGT). A variable turbine geometry allows for the change of the turbines' aspect ratio. The aspect ratio can be manipulated in several different ways (the interested reader is referred to [10]). However, most systems use moving vans to decrease the inlet area of the turbine in engine operating points with a low engine speed. The velocity of the mass flow increases and therefore, the available kinetic energy at low engine speed engine operating points. For high mass flow rate engine operating points, the turbine inlet area is increased to inhibit choking effects. However, exhaust gas temperatures of gasoline engines exceed 1000 °C, which prevents the current state of the art VGT from being applied in these engines.

Using an electrically driven compressor would decouple the increase of boost pressure and the mass flow rate of the turbine. This allows a variable and easily controllable torque build-up at low engine speed [3]. However, this approach requires the supply of electrical energy.

The changes in the turbocharger system that directly affect the turbine's efficiency map like changing the size of the turbine and a variable turbine geometry involve disadvantages. The injection of additional air into the exhaust manifold is a possibility to increase the mass flow at the turbine instead of changing its efficiency map. Such a system is developed, for example, by Volvo for the usage in diesel engines, as shown in [11]. This allows the increase of the turbine speed in low engine speed areas. Besides the additional necessary components, an application in a gasoline engine increases the air-fuel ratio and therefore, the three-way catalyst's conversion rate is limited. Rich combustion inside the cylinder allows a stoichiometric operation of the engine. The temperature increase due to the post-oxidation effects are investigated in [29] and [20]. However, the aim of these investigations differs from this work since the focus lies on the catalyst heating effects instead of the shift in low-end torque. The additional components do not only increase the complexity of the system but also require additional efforts in the engine packaging.

A second approach to increase the turbine's mass flow rate is using scavenging air [7]. An overlap of the exhaust valve and intake valve combined with a positive pressure gradient leads to an airflow from the intake manifold directly to the exhaust manifold. The increased mass flow rate is comparable to the secondary air injection approach but without using additional necessary components. As described above, this approach also requires rich combustion in the cylinder to obtain a stoichiometric air-fuel ratio in the engine. Since the scavenging air and the incomplete burnt exhaust gas are spatial separated, the mixing inside the exhaust manifold is crucial to increase the exhaust gas enthalpy using the post-oxidation effect [32]. The interaction of scavenging air and incomplete burnt exhaust gas demands a detailed investigation based on 3D-CFD results coupled with reaction kinetics. A rating of the mixing effects, as well as the thermochemical interaction of the included chemical species, is necessary to evaluate how the post-oxidation process takes place. The operation of an engine, including scavenging and post-oxidation effects, as a tool of low-end torque shift, requires the possibility to develop an adequate engine operating strategy. However, no reliable 0D/1D simulation approach is available that is based on detailed investigations on the post-oxidation process and its related effects. In [28], a simplified 1D simulation approach is shown. Since this approach aims at a potential assessment, it does not show an investigation on the fundamentals of the post-oxidation or does not claim to display the influencing effects of the post-oxidation process. Neither the reaction kinetics is investigated in detail nor the mixing inside the exhaust manifold. A similar approach is discussed in [32] including a similar limitation of the predictive ability.

This work concentrates on developing a 1D post-oxidation model that combines the effects of mixing and reaction kinetics. The chosen approach requires an investigation on the micro-level by means of a detailed investigation of the reaction kinetics and macro-level with the aid of 3D-CFD simulation results, including reaction kinetics.

The simulation of chemical reactions and combustion processes is widely used in the engine development process. The used methodologies are based on thermochemistry fundamentals, including the usage of reaction mechanisms and the chemical equilibrium. In the following, the fundamentals of thermochemistry are discussed to an extent, necessary to follow the thought processes of this work. Subsequent, the fundamentals of a state of the art 0D in-cylinder combustion model is elaborated.

2.1 Fundamentals of Thermochemistry

This chapter provides an overview of the thermochemical fundamentals that are required in the course of this work. At first, the fundamentals on reaction kinetics are discussed. Using these fundamentals, the chemical equilibrium is introduced. Within reaction kinetic simulations reaction mechanisms are used. These mechanisms include specific chemical species and chemical reaction equations, describing the systems behavior. Based on the introduced fundamentals of thermochemistry, the operating principle of reaction mechanisms in the chemkin-II format [18] is discussed.

These fundamentals do not claim to cover the topic of thermochemistry in its entirety. Therefore, the interested reader is referred to [34] and [35].

In equation 2.1 a fictive chemical reaction equation is given, including the fictive species A and B as reactants and the fictive products C and D. The factor ν_i relates to the reaction order ([34], [35]).

$$\nu_A \cdot A + \nu_B \cdot B \xrightarrow{k_f} \nu_C \cdot C + \nu_D \cdot D \qquad 2.1$$

The reaction consumes the species A and B to produce the species C and D. The parameter that describes the rate of this reaction is k_f. The consumption of A can be described by equation 2.2.

$$\frac{d[A]}{dt} = -k_f[A]^{\nu_A}[B]^{\nu_B} \tag{2.2}$$

The expression [A] represents the concentration of A and is given in mole per m^3. The higher k_f, the faster the decrease of [A]. The decrease of [A] is also proportional to [B] and [A] itself. The higher the reactant concentrations, the more likely is an interaction of these. The reaction orders ν_i describe the change in reaction rate with changing involving species concentration. The overall reaction order is (ν_A + ν_B). The reaction rate constant k_f is proportional to the temperature and its dimension changes with a changing overall reaction order. The empirical modified Arrhenius approach describes the reaction rates temperature dependency as follows:

$$k = AT^b e^{\frac{E_A}{R_u T}} \tag{2.3}$$

While E_A describes the activation energy (the energy barrier, which needs to be overcome), the pre-exponent multiplier A has different meanings for uni-, bi- and trimolecular reactions. Based on the bimolecular reaction equation 2.1, A corresponds to the product of the collision rate of molecules and the probability of the reaction [35]. The exponent b takes the increased reactivity with increasing temperature into account. The reaction of equation 2.1 not only includes the formation of products but also the dissipation of these products as soon as the temperature is high enough. As a result, the reaction also proceeds in the opposite direction. The corresponding reaction rate is given with the reverse reaction rate k_r as a counterpart to the until now used forward reaction rate k_f. The resulting complete reaction equation is

$$\nu_A \cdot A + \nu_B \cdot B \underset{k_r}{\overset{k_f}{\rightleftharpoons}} \nu_C \cdot C + \nu_D \cdot D \tag{2.4}$$

In consequence, the consumption of species A changes since depending on k_r, A is also produced. The following equation expresses this context.

$$\frac{d[A]}{dt} = -k_f[A]^{\nu_A}[B]^{\nu_B} + k_r[C]^{\nu_C}[D]^{\nu_D} \qquad 2.5$$

A high difference in the consumption rate of A $(-k_f[A]^{\nu_A}[B]^{\nu_B})$ and the production rate of A $(+k_r[C]^{\nu_C}[D]^{\nu_D})$ causes a significant change in [A] over time. A minor difference causes a lower change in [A] over time. Consequently, the consumption rate and the production rate are approaching each other until they are equal. This condition is called chemical equilibrium, and the concentration [A] is constant:

$$\frac{d[A]}{dt} = -k_f[A]^{\nu_A}[B]^{\nu_B} + k_r[C]^{\nu_C}[D]^{\nu_D} = 0 \qquad 2.6$$

This equation can be transformed as follows:

$$\frac{k_f}{k_r} = \frac{[C]^{\nu_C}[D]^{\nu_D}}{[A]^{\nu_A}[B]^{\nu_B}} = K_C \qquad 2.7$$

The equilibrium constant K_C depends on the concentrations of the contributing species. According to [34], the equilibrium constant can also be expressed concerning the partial pressures of each involved species:

$$K_P = \frac{p_C^{\nu_C} p_D^{\nu_D}}{p_A^{\nu_A} p_B^{\nu_B}} = K_C(R_u T)^{\nu_C+\nu_D-\nu_A-\nu_B} \qquad 2.8$$

The equilibrium constant K_P can be calculated based on the Gibbs free energy of each involved species. The Gibbs free energy of each species can be calculated based on the species individual molar enthalpy and molar entropy. These values can be taken from tables. The interested reader is referred to [34] and [35].

With this information, an equation system can determine the chemical equilibrium concentration of a chemical reaction. The size of the equation system is proportional to the considered amount of species. The chemical equilibrium is the resulting species concentration within a system that is given an infinite amount of time. Since the reaction rates are not infinitely high, the actual concentrations approach the chemical equilibrium but depend on the available time. In a system with a relatively high temperature, the reaction rates are also quite high. Therefore, the species concentration does approach the chemical equilibrium in a lower time. With decreasing temperatures, the reaction rates are getting lower and therefore, the necessary amount of time to reach the chemical equilibrium state increases.

2.2 Operating Principle of Reaction Mechanisms

Subsequent, reaction mechanisms are used to display the reaction kinetics of several chemical systems in a simulative way. This section gives an overview on their operating principle.

The overall oxidation process of fuel and oxidizers can be expressed by a reaction equation like equation 2.1. To give a more specific example, the oxidation process reaction equation of H_2 can be express as follows:

$$2H_2 + O_2 \longrightarrow 2H_2O \qquad 2.9$$

This global reaction equation can display the overall oxidation process of H_2. However, it is improbable that two H_2 molecules approach an O_2 molecule simultaneously to form two H_2O molecules. This would require breaking several chemical bonds while forming several new chemical bonds subsequently. The reaction equation explains the overall oxidation process, but it does not point out the single steps involved in the oxidation process [34].

The global reaction equation 2.9 combines many elementary reactions that describe the actual kinetic chemistry, including several intermediate species. The collision of H_2 and O_2 is more likely to form HO_2 and H as intermediate species (equation 2.10) instead of H_2O since the collision of only two molecules is more likely to happen. Also, the formation of HO_2 requires only one chemical bond to be broken and one to be formed [34].

$$H_2 + O_2 \longrightarrow HO_2 + H \qquad 2.10$$

The intermediate species that are produced by an elementary reaction do influence the oxidation process in different ways. An intermediate species with unpaired electrons is very likely to create new chemicals bonds when approaching another species and is considered very reactive. These species are known as radicals or free radicals. An elementary reaction that produces a radical without a radical being involved as the reactant is a chain-initiation reaction. An elementary reaction that produces one radical with one radical being involved as the reactant is a chain-propagating reaction. They do not speed up the reaction by increasing the amount of free radicals but also they are not slow down the reaction significantly since they do not reduce the overall radical quantity. Consequently, elementary reactions that increase the overall amount of radicals by producing two radicals with one radical being involved are chain-branching reactions. Chain-terminating reactions reduce the amount of radicals by producing only one out of two radicals or by the generation of no radical at all with minimum one radical being involved as reactant [34].

Many oxidation processes rely on the presence of these free radicals. The oxidation of the species that are considered the fuel often relies on the interaction with free radicals. As mentioned before, the reaction rates of chemical reactions are proportional to the system's temperature. An increasing temperature increases the reaction rates. A low system temperature results in a slow formation of free radicals by chain-initiation and chain-branching reactions. With an example in mind as displayed in equation 2.2, where A is considered the fuel and B the necessary free radical, it is observable that a low k_f due to low temperatures needs to be compensated with a high concentration of the radical B in order to achieve a high consumption rate of A.

The production of B with relatively low temperatures requires a specific amount of time. As soon as the interaction of A and B gets more likely, the oxidation reaction – as displayed in equation 2.1 – occurs in a higher frequency. Each interaction releases a specific amount of heat and the system's temperature increases. With increasing temperature, the reaction rates of both, the oxidation reaction and the chain-initiation/branching reactions increase. The result is a self-reinforcing effect called ignition. The necessary time span of radical formation to start the ignition is the chemical ignition delay [34], [35].

The global reaction equation 2.9 covers the overall oxidation process of H_2 that needs more than 20 elementary reactions to be described in detail [34]. A detailed investigation of the hydrogen oxidation process requires considering all or at least the most important of these elementary reactions.

In this work, the investigation of oxidation processes is made by usage of reaction mechanisms. Reaction mechanisms contain the necessary elementary reaction equations to describe the oxidation process of a specific set of species within a validated range of boundary conditions.

The reaction mechanisms used in this study (e.g. [4] and [12]) contain information of the involved species and the required elementary reaction equations. The forward reaction rate k_f of each elementary reaction is given by defined parameters of the modified Arrhenius approach (equation 2.3). This allows the calculation of k_f for each reaction equation concerning the current temperature.

Additional thermodynamic quantities are represented with the NASA polynomials (equations 2.11 - 2.13).

$$\frac{c_p}{R} = a_1 + a_2T + a_3T^2 + a_4T^3 + a_5T^4 \tag{2.11}$$

$$\frac{H^0}{RT} = a_1 + \frac{a_2}{2}T + \frac{a_3}{3}T^2 + \frac{a_4}{4}T^3 + \frac{a_5}{5}T^4 + \frac{a_6}{T} \tag{2.12}$$

$$\frac{S^0}{R} = a_1 ln(T) + a_2T + \frac{a_3}{2}T^2 + \frac{a_4}{3}T^3 + \frac{a_5}{4}T^4 + a_7 \tag{2.13}$$

For each species, the reaction mechanism includes the parameters a_1 – a_7. Commonly, these parameters are set in a different way for two temperature ranges. The NASA polynomials are used to calculate the theoretical chemical equilibrium constants K_P or K_C. In combination with equation 2.7, k_r is determined. This is necessary since all involved reaction equations are reversible [18].

Available reaction mechanisms are the outcome of elaborate and complex development activity. Reaction mechanisms that cover the oxidation process of long-chain hydrocarbons include several thousand elementary reactions and hundreds of individual species. With more complex reaction mechanisms, the necessary computation time of a reaction kinetics investigation increases. A reduction of these mechanisms is made by elaboration, which reaction equations are negligible for the investigated boundary condition range. The resulting reduced reaction mechanism can display the same oxidation process within a specific boundary conditions range but with fewer reaction equations and less computation time. Possible mechanism reduction methodologies are described in [22] and [25].

2.3 0D/1D Simulation Domain

2.3.1 1D Fluid Mechanics

A 0D/1D full engine model is a combination of 1D simulation and 0D simulation approaches. A 0D/1D simulation model contains all relevant flow components of an engine. These elements are, in contrast to 3D-CFD, discretized only in axial direction.

The flow field is described by the Navier-Stokes equations that include the mathematical expression of conservation of mass, conservation of momentum and conservation of energy. A complete discretization in all three dimensions would require an overall equation number of five (conservation of mass, conservation of momentum in all three dimensions and conservation of energy). Each of these equations needs to be solved several times in each discretized time step until they reach convergence. This needs to be done in each discretization volume of the simulated mesh.

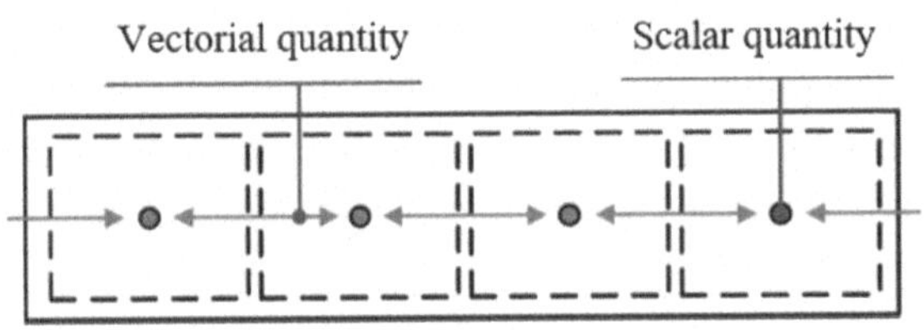

Figure 2.1: Schematic illustration of the 1D approach discretization [30].

The one dimensional discretization allows the reduction of necessary equations that need to be solved. Since the underlying solver needs to cover each discretized part of the flow model on its own, a lower amount of discretization section saves additional computation time [21]. A schematic illustration of a pipe discretized by means of 1D discretization is displayed in figure 2.1. The markers symbolize the scalar values inside each discretization section. For example the local temperature that is a scalar value for the whole discretization section and therefore is averaged in this section. The arrows are the vectorial values. In combination, these both value types describe the current state of each discretization section (scalar values) and the influence of each neighboring section (vectorial values).

The flow field inside the 1D engine flow components is characterized by fast travelling pressure waves. Hence, important indicators of the engine simulation model, for example the cylinder charge, are influenced by the pressure waves, the choice of flow field solver is important. An explicit solver calculates the state of each discretization section based on its own state (scalar values) and also based on the state of its neighboring sections (vectorial values). The state of each subsection is calculated individually. An implicit solver needs to set up a system of equations that include the solution of every involved subsection. Then the system of equation is solved to generate the current state of each involved subsection. This methodology requires more effort for each time step but due to its characteristics, a greater time step can be chosen. As the time step increases, the disadvantage of the more complex calculation method will eventually be eliminated. Since the flow field needs the pressure waves to be resolved, a rather small time step is necessary. The implicit solver cannot use its advantages here and therefore the explicit solver is chosen.

The choice of discretization length (describes the size of a 1D discretized subsection) does not only affect the computation time by means of changing amount of subsection but also because of its influence on the maximum size of the simulations time step. With a given discretization length Δx, the corresponding time step Δt cannot be chosen arbitrarily. In terms of fluid dynamics, the maximum time step is set by the speed of sound. The speed of sound describes the information's speed of propagation inside a flow field. Within a simulation approach including a flow field, the numerical information's speed of propagation is not allowed to be higher than the speed of sound. This would lead to different flow characteristic since the propagation of pressure waves depends highly on the speed of sound as characteristic speed of information [24]. The maximum numerical speed of information is set by the choice of discretization length and time step: $\Delta x/\Delta t$.

To ensure a stable system, the Courant-Friedrichs-Lewy condition (CFL) is used according to equation 2.14.

$$CFL = (|u_{fluid}| + u_{sound})\frac{\Delta t}{\Delta x} \leq CFL_{max} \tag{2.14}$$

Due to the choice of an explicit solver, CFL_{max} equals 1 (or even 0.8 for a more stable solver behavior) u_{fluid} represents the local fluid velocity and usound the speed of sound. The choice of a halved discretization length requires therefore also a halved time step. The resulting computation time is increased by the factor 4 [21], [24].

Since the main flow direction of the fluid inside the engine is covered by only an axial discretization, the relevant influences can be covered and the 1D flow model approach is a valid methodology. However, some part of the engine show a highly three dimensional flow behavior, for example the cylinders and the turbocharger. These parts are not resolved by a 1D flow mechanics but with a 0D approach. In the following the 0D quasi-dimensional combustion modelling approach of the cylinder combustion process is explained.

2.3.2 Quasi-Dimensional Combustion Modelling of a Spark-Ignition Engine

This subchapter gives an impression on the quasi-dimensional modelling of the combustion process inside a spark ignition (SI) engine. This combustion model is used in a later stage of the work to determine each cylinders emission composition and the corresponding temperatures and pressure. This model approach is a state of the art combustion model and is therefore widely used in 0D/1D engine simulation investigation.

The quasi-dimensional combustion model covers the high-pressure part of the in-cylinder combustion. The high-pressure part is the time span of the combustion process, beginning with the intake valves' closing event and ending with the exhaust valves' opening event. The combustion chamber can be described by use of the following thermodynamic definitions. The first step is to define the area of interest as a thermodynamic system. A thermodynamic system is surrounded by the system boundaries, whose characteristics depend on the system type. In this particular case, the investigation will be made during the high-pressure part and therefore with closed valves. In order to describe these characteristics, the chosen system needs to be a closed system. Closed thermodynamic systems do not allow any kind of matter to pass the boundaries but are permeable to energy. This attribute is necessary to display heat fluxes due to wall heat losses and the adding or removal of work from the system (piston movement) [36]. The closed thermodynamic system, representing the combustion chamber, can then be further divided into two subzones. These subzones are a homogenous subpart of the system. The temperature and the present species is constant for the whole system and therefore for both zones. The different subzones inside the system must not overlap each other. The whole system's pressure is constant over the whole area, and therefore the pressure of each subzone is equal. One zone represents the unburnt part of the combustion chamber while the second zone represents the burnt part. They are separated by the flame front, which belongs to the burnt zone from a thermodynamic point of view. On the one hand, the unburnt zone contains a homogenous mixture of air, fuel, and residual exhaust gas. The burnt zone, on the other hand, contains burnt exhaust gas ([14] and [15]). Figure 2.2 shows a schematic illustration of the thermodynamic system "combustion chamber" and its two subzones.

The flame front penetrates the unburnt zone and its speed is globally seen and defined as follows:

$$u_e = u_{turb} + u_L \qquad 2.15$$

The flame front speed is the sum of the laminar flame speed u_L and the isotropic turbulence speed u_{turb} [14]. Hence, the mass entering the flame zone is defined as

$$\frac{dm_e}{dt} = \rho_{ub} A_{Fl} u_e \qquad 2.16$$

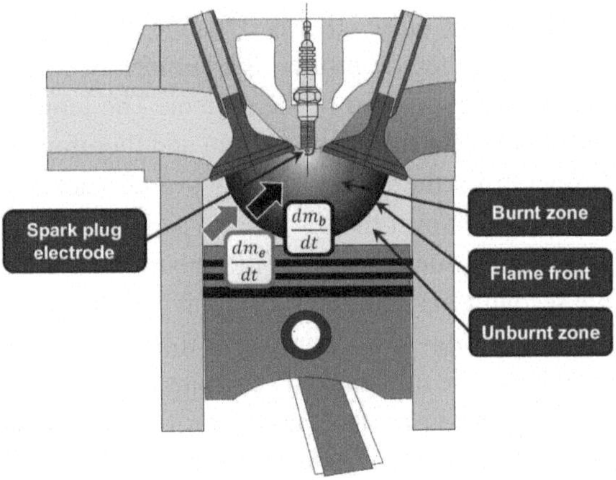

Figure 2.2: Schematic illustration of the two-zone quasi-dimensional combustion model [8].

ρ_{ub} represents the density of the unburnt zone and A_{Fl} is the surface of the flame. The mass flow into the burnt zone is defined as

$$\frac{dm_b}{dt} = -\frac{dm_{ub}}{dt} = \frac{m_F}{\tau_L} \qquad 2.17$$

with m_F as the mass of the flame zone and τ_L as the characteristic burn time. Further details on the flame propagation can be found in [14]. The focus in this work lies on the species inside the burnt zone due to the combustion process.

As described in chapter 2, the species composition in a chemical equilibrium depends on the pressure, temperature, current species ratio, and the equilibrium constants of the present species. A species calculation approach for eleven species is presented in [14]. This approach uses the hydrogen-carbon ratio of the used gasoline fuel surrogate to calculate. As a result, it is possible to calculate the species concentration inside the burnt zone for each time step. The species concentration is held constant as soon as the temperature falls below a certain threshold temperature to display the reaction rate's temperature dependency.

3 Investigation on the Post-Oxidation Effect

This work focuses on the oxidation process of CO and H_2 inside the exhaust manifold – to be more precisely in between of the exhaust valves and the turbine. This phenomenon occurs as soon as the torque-increasing effect of scavenging is used while the engine's overall air-fuel ratio is held on a constant stoichiometric level. The 1D simulation domain is of great importance during an engine's development process and its operating strategy. The additional heat release close to the turbine might have an important influence on the turbine operating point and therefore on the entire engine operation. To cover this effect during the engine development process, a reliable post-oxidation model is necessary. In the following, the post-oxidation effect is investigated in great detail.

The aim is to determine the most important parameters influencing post-oxidation. First, the reaction kinetic correlations of CO and H_2 oxidation are illuminated on the micro-level. This is followed by investigations on the macro-level, based on 3D-CFD simulation results analyzed in detail in this second section concerning post-oxidation. The evaluated 3D-CFD simulation results are taken from [1].

The investigation is based on the research project documented in [1]. The project includes a test-bench investigation and 3D-CFD simulations of this very test-bench engine. The test-bench engine is a turbocharged four-cylinder inline DISI engine. The volume capacity is 1618 cm^3 with a compression ratio of 10.5:1. The rated power is 140 kW at 5600 rpm. The engine model includes a 10 cm adaptor downstream of each of the four cylinders. These adaptors are necessary to include measurement devices. The test-bench engine's valve timing allows the usage of scavenging air to investigate post-oxidation effects.

The 3D-CFD simulation of this test-bench engine are validated in terms of combustion calibration, exhaust gas temperature and exhaust gas composition (briefly discussed in 3.2.1 and detailed included in [1]. The validated 3D-CFD simulations allow a deep insight into the post-oxidation process.

J. Przewlocki, *Simulative Investigation of Post-Oxidation in the Exhaust Manifold of SI Engines*, Wissenschaftliche Reihe Fahrzeugtechnik Universität Stuttgart,
https://doi.org/10.1007/978-3-658-36378-9_3

3.1 Reaction Kinetic Investigation

This chapter deals with reaction kinetic investigations of the post-oxidation process at boundary conditions related to an SI engine's exhaust manifold. The reaction kinetic investigation is carried out with the open-source reaction kinetics software Cantera [13].

3.1.1 0D Reaction Kinetics Simulation Setup

The simulation methodology contains a 0D reactor with constant pressure and an ideal gas simplification. The reactors content is a gaseous fluid, whose temperature, pressure and composition can be individually set. A reaction mechanism describes the reaction kinetics behavior of the content of the reactor. The operating principle of a reaction mechanism is described in chapter 2.1.

The reaction kinetics simulation setup is split into two steps. The first step is the generation of boundary conditions by means of species composition. The second step is the combustion of these species after being mixed up with a specific amount of air.

To perform reaction kinetics investigation focusing on the post-oxidation phenomenon, the boundary condition regarding incomplete burnt combustion products mixed up with unburned air needs to be set. In the following, a simplified approach is used to define incomplete burnt combustion products. A surrogate gasoline fuel (in this case 90 $\%mol$ iso-octane and 10 $\%mol$ n-heptane) is mixed up with air (21 $\%mol$ oxygen and 79 $\%mol$ nitrogen) according to the in each case used air-fuel ratio. With the usage of chemical equilibrium (in detail described in chapter 2.1 at a certain temperature and pressure, the composition of a gas with a composition similar to exhaust gas can be defined (note: species, which are only present in traces are ignored). After this first step, the surrogate exhaust gas can then be remixed with air to investigate the oxidation behavior of incomplete burnt combustion products under exhaust manifold thermodynamic conditions. Within the subsequent investigation, this methodology is used to generate surrogate exhaust gas.

The post-oxidation phenomenon investigation focuses on engine operating points with an engine speed below $\approx$ 2000 rpm. Therefore, an approximated maximum exhaust gas residence time inside the exhaust manifold of 5 ms is assumed. Hence, the reaction kinetics investigation of the post-oxidation process focuses on a time span of 0 to 5 ms.

Choice of Reaction Mechanism

The reaction kinetics investigation, as well as the 3D-CFD simulations of the exhaust manifold used within the research project on which this work is based [1], require the choice of a reaction mechanism. Several reaction mechanisms are available, and each is developed in order to fit individual requirements. They differ in the available chemical species, the implemented reaction equations, and the definition of these reaction equations regarding the Arrhenius equation parameter (compare chapter 2.2). The choice of a reaction mechanism bases on a trade-off regarding available species, implemented reaction equations and computation time. In addition, each reaction mechanism is developed and validated for a certain boundary condition range and species composition. Therefore, this subchapter is dedicated to comparing different reaction mechanisms, which do suite the requirements regarding necessary species and boundary conditions range used for validation purpose.

The comparison is made by the oxidation of incomplete burnt combustion products (as defined in chapter 3.1.1) under each identical conditions within a 0D constant pressure ideal gas reactor. The reaction mechanism used to define the rich exhaust gas composition (step 1) is not changed to maintain the same conditions for all compared reaction mechanisms (step 2). The mixture's air-fuel ratio, used to produce the incomplete burnt combustion products, equals 0.8 ($\lambda_{Combustion} = 0.8$). The air-fuel ratio of the mixture of the second step (incomplete burnt combustion products and air) mixture equals one ($\lambda_{Post-Oxidation} = 1$). Table 3.1 lists the gas compositions, which are used in step 1 and step 2. Table 3.2 gives an overview of the used reaction mechanisms.

Table 3.1: Approximate boundary condition gas composition of step 1 and step 2.

	Iso-Octane	N-Heptane	H_2O	CO_2	CO	H_2	O_2	N_2
Mixture Step 1 [-]	0.9	0.1	-	-	-	-	9.88	37.15
Mixture Step 2 [-]	-	-	0.101	0.082	0.034	0.041	0.037	0.704

Table 3.2: Reaction mechanism overview.

Name	Amount of Species	Reaction Equations	Reference
Reaction mechanism 1	156	3465	[31]
Reaction mechanism 2	14	33	[12], [27]
Reaction mechanism 3	32	173	[9], [27]
Reaction mechanism 4	487	2081	[4]
Reaction mechanism 5	323	2469	[23]
Reaction mechanism 6	15	48	[19]

Figure 3.1 shows the temperature of the 0D reactor within a time span of 2 ms. The temperature increase is due to the chemical heat release during the CO and H_2 oxidation process. It can be observed that the used reaction mechanisms differ in ignition delay and temperature gradient. The deviation of the reaction mechanism results is rather small, especially regarding the differences in included species and reaction equation quantity.

The choice of a reaction mechanism is based on the requirements of both investigation types, 0D reaction kinetics and 3D-CFD simulation. It is intended to use the same reaction mechanism to achieve reliable comparability between these two simulation domains.

The reaction mechanisms of choice are reaction mechanism 1, 2 and 3. These reaction mechanisms originate from the same detailed mechanism that includes

451 species and 17848 reaction equations. A reduction of this mechanism in order to generate reaction mechanism 1 can be found in [31] while the further reduction can be found in [12] and [27]. Reaction mechanism 3 contains the same $CO/H_2/O_2$ system as reaction mechanism 2 and includes reaction paths for the NO_x investigation [9]. The choice of these three mechanisms allows the fast computation of a $CO/H_2/O_2$ system inside the exhaust manifold, considering more complex hydrocarbons and NO_x with a maximum of comparability with very good comparability.

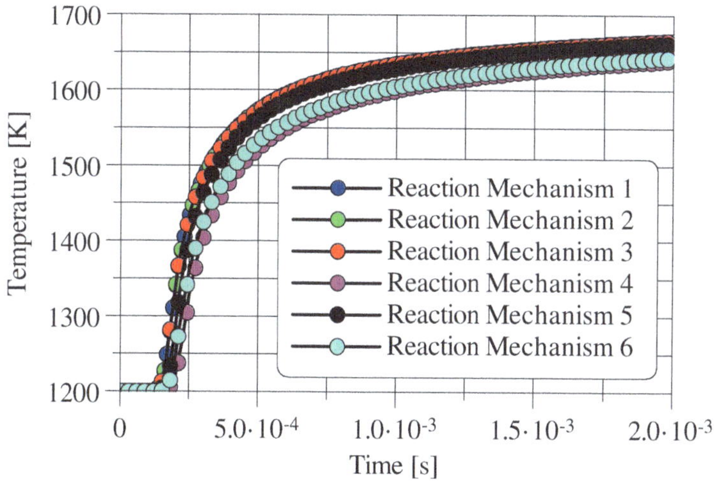

Figure 3.1: Comparison of reaction mechanisms with an equal gas composition and equal temperature and pressure boundary conditions.

3.1.2 Investigation on Carbon Monoxide and Hydrogen Oxidation

This chapter concentrates on the reaction kinetics of the CO and H_2 oxidation. Its target is to give background information on the chemical processes, which occur during the post-oxidation process under boundary conditions comparable to the exhaust manifold boundary conditions. The subsequent investigation is again carried out with the reaction kinetics software Cantera [13]. The simulation setup is similar to the already described two-step simulation setup.

The chosen starting temperature is 1200 K, p = 1 bar and reaction mechanism 2 is implemented (compare Table 3.2).

Figure 3.2 illustrates the change in temperature, net progress rate of reaction equations and species mole fraction during a time span of 2 ms for a mixture of incomplete burnt combustion products and air. The net rate of progress is the difference between the forward rate of progress and the reverse rate of progress of a specific reaction equation. The rate describes the importance and direction of a chemical reaction equation. A positive value implies a greater forward rate of progress, while a negative value stands for a larger negative rate of progress [13].

The net rates of progress in Figure 3.2 are the most relevant for this oxidation progress. It can be observed that reaction equation 3.1 is the most important reaction equation of the CO oxidation process.

$$CO + OH \rightleftharpoons CO_2 + H \quad 3.1$$

$$CO + \frac{1}{2}O_2 \rightleftharpoons CO_2 \quad 3.2$$

On the other hand, the reaction equation 3.2 can be neglected because of its low reaction speed. This finding is well known and can be found in a wide range of literature, for example, in [34] and [35].

The fact that the oxidation of CO requires an OH-radical leads to the fact that the production of OH also needs to be displayed by a reaction mechanism if the target is to describe the CO oxidation within the 0D and the 3D-CFD simulation domain. This radical production is described by the $O_2 - H_2$ system, which is discussed briefly in the following.

Some of the most relevant reaction equations of the O_2/H_2 system for this specific case are according to Figure 3.2:

$$H + O_2 \rightleftharpoons O + OH \quad 3.3$$

$$H_2 + O \rightleftharpoons H + OH \quad 3.4$$

$$H + H_2O \rightleftharpoons OH + H_2 \quad 3.5$$

$$H + OH(+M) \rightleftharpoons H_2O(+M) \quad 3.6$$

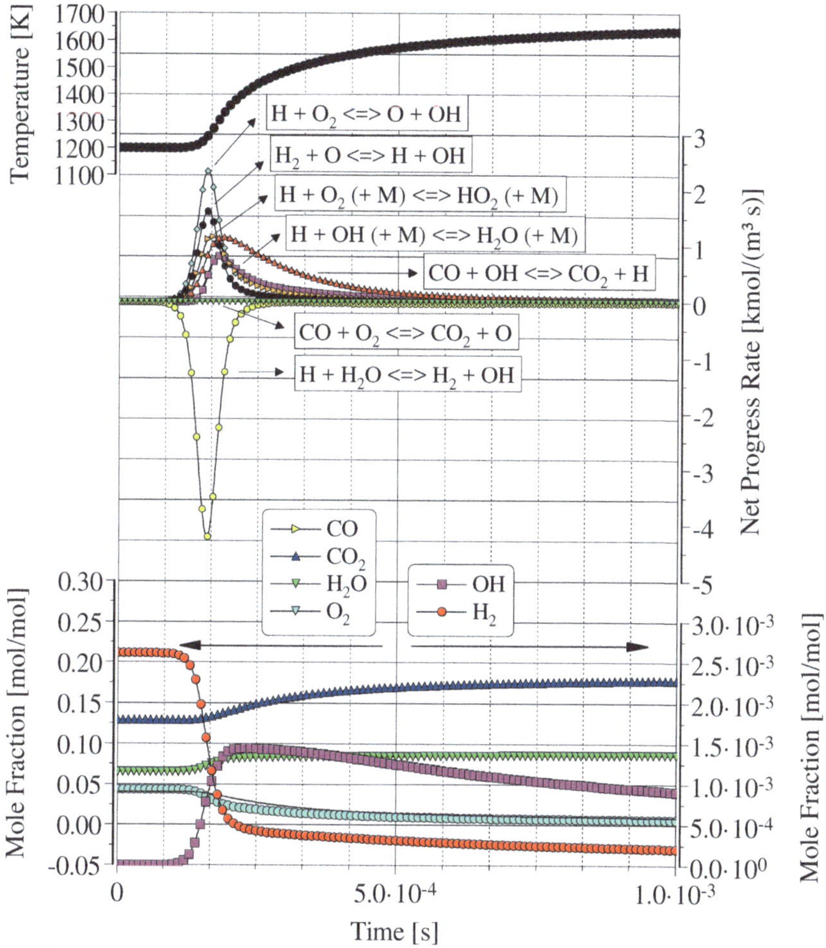

Figure 3.2: Changes in temperature, net rate of progress and species composition during a 0D reaction kinetics investigation of the CO/H_2 combustion process.

Reaction equation 3.3 is a chain branching step (formation of two radicals out of one radical) while reaction equations 3.4 and 3.5 are chain transfer steps (constant amount of free valences). Reaction equation 3.6 decreases the amount of radicals and is, therefore, a chain-terminating step [34]. The negative net rate of progress of reaction 3.5 in Figure 3.2 shows that this reaction is responsible for the net H_2O formation in this oxidation process (negative values imply production of the species on the left side of the reaction equation).

Beside the reaction equation 3.5, several other reaction equations do in fact produce H_2O during the oxidation process. Some of these show a higher forward rate of progress, but at the same time, they do also show a high reverse rate of progress. This results in a high H_2O production rate but also in a high H_2O consumption rate. Therefore, the net production rate of H_2O is lower than the net production rate of reaction equation 3.5.

In order to get a more detailed insight into the reaction pathways during the oxidation process, a reaction path diagram is used. Figure 3.3 shows a reaction path diagram within the same simulation setup as used in Figure 3.2. It shows the fluxes of the element O by the usage of arrows, whose width are proportional to the net production rate. This illustration points out the importance of the OH-radical formation for the whole oxidation process. The entire production of CO_2 and H_2O depends on this radical and therefore the production of OH can be seen as some kind of bottleneck of this oxidation process. The reaction path of the single-atom O to CO_2 underlines the subordinate role of reaction equation 3.2 again.

Beside the temperature and the net rates of progress, Figure 3.2 also shows the mole fraction of the most relevant species. Appropriate to the findings so far, the mole fraction of OH increases while the mole fraction of H_2 decreases. This first change in mole fraction occurs before the temperature increases noticeably. With an increasing amount of present OH, the oxidation of CO starts. This delay of temperature increase due to chemical processes (like the OH-radical production) is called chemical ignition delay and discussed in chapter 3.1. The oxidation of CO releases a great amount of heat, and therefore, the rise in temperature starts appropriately. Reaction equation 3.6 also starts to accelerate with an increasing amount of OH available. The relatively high lower heating value of H_2 leads to the assumption that this reaction equation also releases a noticeable amount of energy during the post-oxidation process in the exhaust manifold.

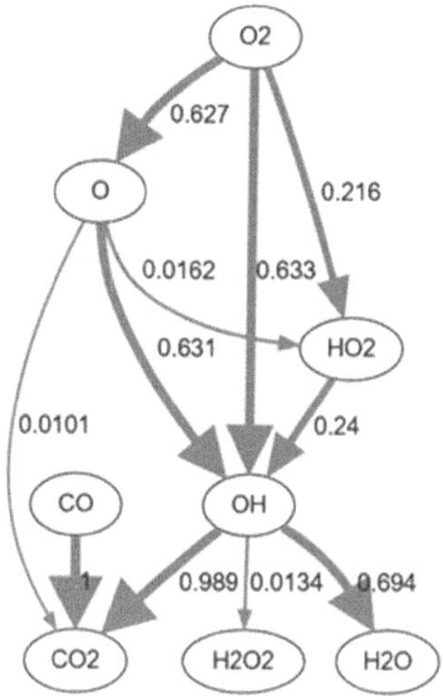

Figure 3.3: Reaction path diagram regarding the element O with a starting temperature of 1200 K and p = 1 bar.

Besides its role as an energy supplier for the post-oxidation process, H_2 also plays a key role in the oxidation process of CO since it is jointly responsible for the OH-radical formation. Without present OH, the oxidation process of CO is mainly driven by the reaction equation 3.2. This prolongs the oxidation process significantly [34].

Dependency of the Carbon Monoxide and Hydrogen Oxidation Process on Temperature and Pressure

As mentioned in chapter 3.1, the reaction rates depend on the temperature and the pressure of the investigated system. The oxidation process inside the exhaust manifold does have a limited time span to appear (as mentioned at the beginning of this chapter). As soon as the oxidation process of CO starts, the temperature increases rapidly, and so does the reaction rates. Thus, this subchapter's focus lies on the chemical ignition delay of the mixture and its dependency on the temperature and pressure.

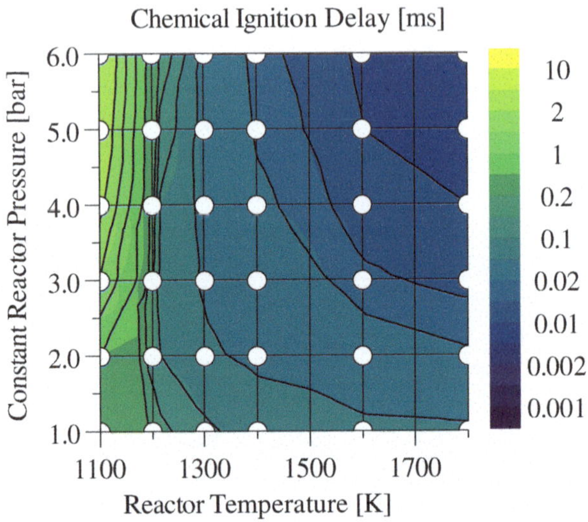

Figure 3.4: Reaction kinetic investigation results on the chemical ignition delay with respect to changing temperature and constant pressure boundary conditions.

In this work, the chemical ignition delay is defined as the time necessary to increase the temperature of the mixture by 100 K above the starting conditions. Figure 3.4 displays the chemical ignition delay for a temperature span of 1000 K up to 2000 K and a pressure span of 1 bar up to 6 bar.

The chemical ignition delay decreases with increasing temperature. This behavior shows the high dependence on temperature of the radical production rate. With a temperature above 1600 K and a pressure below 2 bar, the ignition delay does not speed up significantly but remains on a high level of approximately 0.02 – 0.05 ms. To decrease the chemical ignition delay significantly, an increase in pressure is necessary.

Within the temperature range of 1400 K to 1600 K, the pressure variation shows a clear effect. The chemical ignition delay decreases with increasing pressure. For temperatures below 1400 K and above $\approx$ 1200 K, the increasing pressure has a negligible effect on the chemical ignition delay. An increase of pressure for temperatures below ≈ 1200 K shows an increase in chemical ignition delay.

Figure 3.5 shows the temperature increase for a starting temperature of 1200 K and a pressure span of 1 bar up to 5 bar. With increasing pressure, the chemical ignition delay decreases until a pressure of 3 bar is reached, then the tendency turns around, and the chemical ignition delay increases with increasing pressure. In the following, this behavior is investigated and discussed in detail.

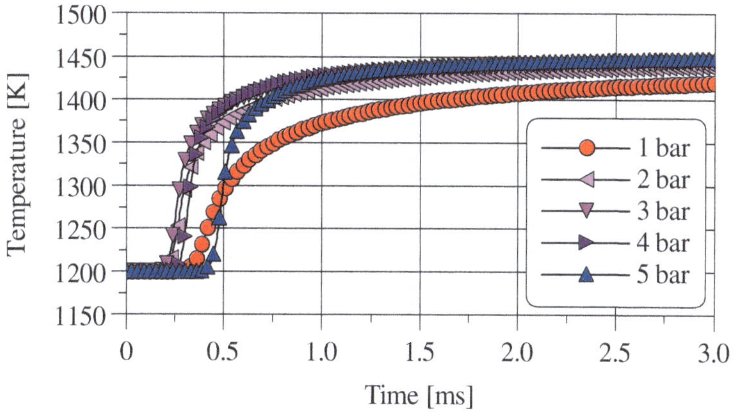

Figure 3.5: Differences in the system's reactivity due to changing constant pressure boundary conditions.

The colder the mixture of incomplete burnt combustion products gets, the more dominant is the production of comparatively unreactive HO_2-radicals instead of OH-radicals [6]. This behavior is displayed in Figure 3.6. Each curve displays the ratio of OH to HO_2 within the same time span, as displayed in Figure 3.5. The curves are each marked (black arrow) at the time step at which the temperature increases by 2 K. By doing this, it can be observed that the mixture needs a certain amount of OH to start the temperature rise. Since the CO oxidation rate depends on the OH concentration, this behavior is not surprising, and the underlying theory is described in chapter 3.1. As soon as a certain temperature is reached (the exact value depends on pressure), the oxidation process increases and a self-reinforcing effect starts.

The amount of *OH*, which is necessary to start this self-reinforcing effect, decreases with increasing pressure. A possible and vivid explanation for this effect is that the reaction rate is dependent on the reactant concentration (discussed in 3.1). The concentration is defined as the amount of a certain species (mole) per volume. This definition displays the increased probability of a species interaction with an increasing amount of involved molecules. The higher the pressure, the more likely is the interaction of *CO* and *OH* since more molecules are present in a specific volume. Thus, the necessary amount of *OH* to start the *CO* oxidation and the temperature increase for higher pressures is lower.

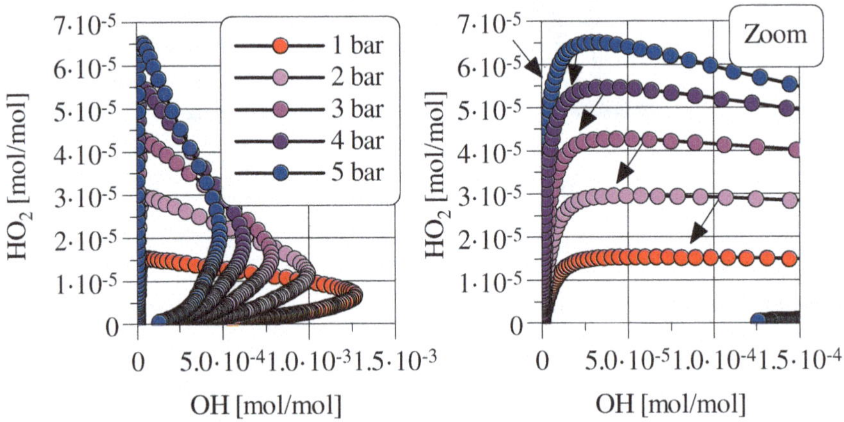

Figure 3.6: Comparison of HO_2 and OH production with different boundary conditions. Left diagram: zoomed illustration of the right diagram.

With further increasing pressure, this ignition delay decreasing effect seems to slow down and at a certain point (in this example for pressure greater than 3 bar), the ignition delay increases again. A possible reason for this might be the H_2O producing reaction equation 3.7:

$$H + O_2(+M) \rightleftharpoons HO_2(+M) \tag{3.7}$$

This results in a change in HO_2 concentration as displayed in equation 3.8:

$$\frac{d[HO_2]}{dt} = k[O_2]^1[H]^1[M]^1 \tag{3.8}$$

Equation 3.8 implies that the reaction rate is proportional to V^3 since each term of species concentration is in the dimension of mass per volume. The thermal equation of state shows that the volume, on the other hand, is inversely proportional to the pressure.

$$V = \frac{nR_uT}{p} \tag{3.9}$$

Thus, the reaction rate of the reaction equation 3.7 is proportional to p^3. The main reaction equations for the OH production (e.g. reaction equation 3.3 and 3.4) include two instead of three reaction partners, and therefore the reaction rate is proportional to V^2 and p^2. The production of OH and HO_2 are both based on H and O_2, and therefore they compete about these species. As a result of this, the production of OH slows down with an accelerating HO_2 production. Figure 3.7 shows the reaction path diagrams of the time span before the temperature increases above 100 K are displayed for 1 bar (a) and 5 bar (b). It can be seen that the reaction rate of O_2 to HO_2 and OH differs. The amount of O_2, which is converted into HO_2 increases with increasing pressure (Figure 3.7 (b)). The HO_2 then converts to OH. Hence, this path takes an additional step compared to the direct path of O_2 to OH, and the chemical ignition delay prolongs as soon as the HO_2 production increases.

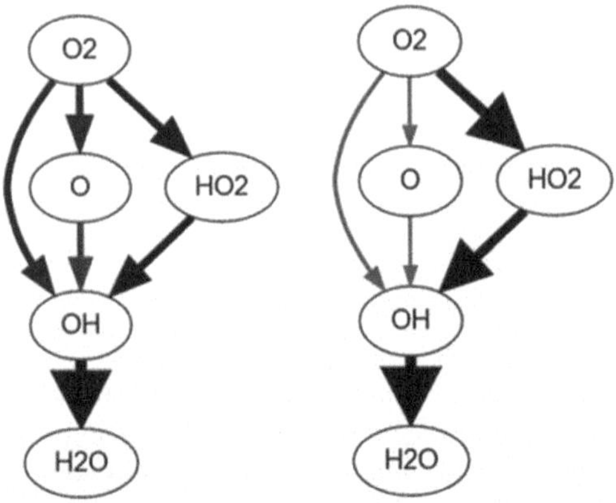

(a) p = 1 bar (b) p = 5 bar

Figure 3.7: Reaction path diagram regarding the element H with a starting temperature of 1200 K and p = 1 bar (a) or p = 5 bar (b).

Influence of Nitric Oxide on the Oxidation Process of Carbon Monoxide and Hydrogen

The chemical ignition delay of the incomplete burnt combustion products and air mixture mainly depends on the production of radicals and as described above depends on the OH/HO_2 ratio. According to [12] and [6], the presence of NO can promote the conversion of HO_2 to OH. Figure 3.7 (b) shows the decelerating effect of the HO_2 production on the CO oxidation due to the inhibition of OH production and the prolonging effect on the reaction path. According to [6], the presence of NO creates an additional reaction path to produce OH out of HO_2 under lean conditions:

$$NO + HO_2 = NO_2 + OH \quad 3.10$$

In combination with reaction equation 3.11, even small amounts of NO can increase the OH production significantly:

$$NO_2 + H = NO + OH \quad 3.11$$

If the combustion takes place under stoichiometric and rich conditions, the presence of NO inhibits the OH production instead. According to [4], the reason is the increasing reaction rate of chain-terminating reactions equations like equation 3.12 due to the increasing available H-radicals.

$$NO + H(+M) = HNO(+M) \quad 3.12$$

The OH production promoting effect has an especially high impact on the CO oxidation as soon as the influence of HO_2 is a major limiting factor. As described above, the importance of HO_2 is the highest for temperatures below 1200 K and high pressures.

Figure 3.8 (a) and (b) show the rise in temperature of an 0D constant pressure reactor according to the setup described in chapter 3.1.1 but with lean post-oxidation conditions ($\lambda_{Post-Oxidation} = 1.2$). The reaction mechanism is changed to reaction mechanism 3 due to its ability to include NO_x species (Table 3.2).

For boundary conditions defined with 1 bar and no available NO, the chemical ignition delay is approximately 1.9 ms. An increasing amount of NO decreases the chemical ignition delay significantly. The presence of 500 ppm NO leads to a more than halved chemical ignition delay.

With a pressure of 3 bar, the chemical ignition delay increases without NO compared to the 1 bar case due to the correlations explained in chapter 3.1.2. The presence of 500 ppm NO decreases the chemical ignition delay significantly. As states above, the presence of NO has a bigger impact on the chemical ignition delay due to the higher amounts of present HO_2.

The chemical ignition delay is in both cases almost similar with 1000 ppm of NO. The rise in temperature starts earlier for the 3 bar case – most likely because of the increased probability of molecule interaction with higher pressures – but the temperature gradient is smaller. Since the conversion of HO_2 to OH does not appear instantly, the reaction equation 3.10 might be the reason for this throttled temperature rise. This throttling effect also occurs in the 1 bar case, but the effect on the temperature rise is lower since the amount of HO_2, which needs to be converted into OH to start the temperature rise, is lower. As soon as the temperature exceeds a certain threshold, the production of HO_2, compared to the production of OH, can be neglected (as described in chapter 3.1.2).

The presence of NO influences the chemical ignition delay of the CO and H_2 oxidation process. The relevance of this influence needs to be further

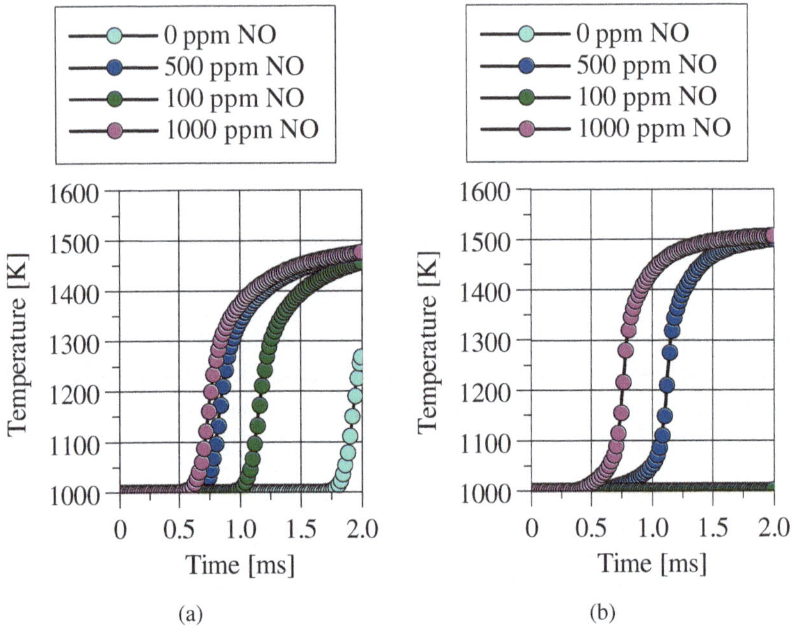

Figure 3.8: 0D reaction kinetic investigation on the effect of NO on the CO oxidation with a starting temperature of 1000 K and p = 1 bar (a) or p = 3 bar (b).

investigated since the local mixture of CO and H_2 and air is crucial for the character of the effect: decreasing or increasing chemical ignition delay. Therefore, a 0D reaction kinetics investigation is able to prove the existence of the effect, but the relevance of the effect on the post-oxidation behavior inside the exhaust manifold needs to be investigated by the usage of 3D-CFD simulations.

3.1.3 Reaction Kinetic Investigation on the 0D Cylinder Raw Emissions

Beside a general investigation on the post-oxidation inside the exhaust manifold, a post-oxidation model development for the fast 1D simulation domain is targeted in this work. The reaction kinetic investigation on the CO and H_2 post-oxidation process shows a sensitive behavior with changing boundary conditions. In addition, the amount of CO and H_2 do directly affect the possible amount of releasable heat during the post-oxidation process. In the subsequent, the current approach of emission calculation of a state of the art 0D combustion model is extended by a reaction kinetics approach to rate the trade-offs regarding accuracy that have to be made for a low computation time.

The current approach on cylinder combustion products calculation is based on the two-zone 0D combustion calculation (as discussed in chapter 2.3.2). The calculation of the species inside the burnt zone is made by usage of a chemical equilibrium calculation (described in chapter 3.1). As discussed in [14], the species concentration inside the burnt zone follows the chemical equilibrium concentration until the temperature of the burnt zone falls below a certain threshold temperature. At this moment, the species concentration is held constant (called "freezing") to take the much slower reaction kinetics at low temperatures into account. Both, the chemical equilibrium, as well as the freezing of the species, are the result of a simplification. This simplification is made to maintain a low computation time, which is necessary for the fast 0D/1D simulation domain.

In the following, the species concentration inside the burnt zone of a two-zone model is recalculated by using a reaction mechanism to investigate whether the simplification of this species calculation methodology can be made or if there is a necessity on a modified species boundary condition calculation methodology. The calculation is made by usage of Cantera [13] and the chosen reaction mechanism for this investigation is reaction mechanism 4.

This investigation is made by the usage of simulation output data. That means that this investigation results do not affect the results of the forerun 1D engine simulation. The target is to recalculate the cylinder raw emissions. A comparison with the emission, originally calculated within the 1D engine model, is then used to evaluate whether a modified simulation calculation approach is necessary.

Input Data

The 0D two-zone combustion model is affected by various influencing parameters delivered by the 1D engine model. Information on, for example, the available amount of fresh air, its temperature and wall heat losses at the cylinder walls is provided by the 1D engine model and its corresponding sub-models. Therefore, several input data are required to recalculate the burnt zones emissions by using reaction kinetics.

Some following boundary conditions must be added as input data since information relies on the cylinder's geometry (for example the wall heat losses during the combustion and the spherical propagation of the flame front inside the cylinder). Other input data relies on the information of the 1D engine model like for example the mass of fresh air inside the cylinders, which is the result of a complex interaction of the intake manifold geometry, the boost pressure provided by the turbine and therefore dependent on the combustion itself.

The required input data is the volume of the burnt zone, the mass of the burnt zone, wall heat losses, the temperature of the unburnt zone, the engine speed, the surrogate fuel composition and the EGR rate:

- The implementation of the burnt zone volume is necessary to take the cylinder geometry into account. With the spherical flame reaches the cylinder walls, the propagation rate changes. This must be taken into account for a comparable emission calculation.
- The mass of the burnt zone can be calculated using the burnt zone volume propagation and the unburnt zone's density. Since the reaction kinetics relies on the burnt zones mass for the temperature and pressure calculation, a direct implementation of the burnt zones mass is the most straightforward way and reduces the risk of error occurrence.
- The heat flux at the cylinder walls directly affects the temperature of both the burnt and unburnt zone during the combustion. The wall heat losses depend on the cylinder geometry and the implemented 1D wall heat model.
- The temperature of the unburnt zone depends beside others on the charge air temperature, the increase in pressure during the compression stroke, the wall heat losses and the fuel evaporation enthalpy.
- The engine speed depends on the chosen engine operating point and is due to the time dependency of reaction kinetic processes a necessity for this modelling approach.

- The surrogate fuel composition determines the species composition of the unburnt zone. Based on this composition, the emissions during the combustion are calculated. These emissions are then located in the burnt zone, and their change in concentration is calculated by the usage of the reaction kinetic software Cantera.
- The change in species composition also depends on the amount of EGR inside the burnt zone. Although the most common species of the EGR mass flow is H_2O and CO_2 that is already burnt, the overall reaction rates and the specific heat capacity are affected by their presence. The amount of EGR depends on the backpressure and valve timings (in case of internal EGR) and therefore must be implemented as a boundary condition.

Modelling of the burnt zone

The simulation of the burnt zone bases on a 0D reactor. The reactor is given the burnt zone volume (input data) for each time step during the calculation time span. With progressing flame front, not only the volume of the burnt zone changes but also the mass inside the burnt zone. The mass flow rate of the flame front into the burnt zone, as described in equation 2.17, is also taken into account as input data.

The reactor has a certain mass and volume for the first time step. As initial species composition, the species composition of the burnt zone out of the two-zone model can be imported but also a pure CO_2 or N_2 as an inertial gas can be chosen. This does not noticeably affect the results due to the very small volume of the burnt zone in the first time step. The initial species composition is necessary to calculate the specific gas constant. This parameter is needed in order to define the actual temperature and pressure of the 0D reactor.

The next time steps lead to an increase in the volume and the mass of the 0D reactor. Both values are input data. Now the species concentration inside the reactor changes due to two different reasons. The first reason is the incoming mass out of the flame front. The second reason is the change in concentration due to the reaction kinetics, which is not only affected by the temperature, pressure and species concentration but also by the available amount of time. This time dependency is the main difference in the chemical equilibrium approach and the usage of this reaction kinetics approach.

The incoming mass flow out of the flame front also needs a yet to be defined species concentration. The definition of this composition relies on a chemical equilibrium calculation. The necessary input parameters are the surrogate fuel composition and the temperature of the unburnt zone. The surrogate fuel composition – containing iso-octane, n-heptane, toluene and possibly ethanol – is already available as an input parameter in the case of this work and the interested reader is referred to [8].

In the first step, the temperature of the flame front needs to be determined. The approximation of choice is the adiabatic isobaric flame temperature. The adiabatic isobaric flame temperature is the maximum achievable temperature of a combustion with constant pressure as a boundary condition. This temperature is reached if the released heat out of a constant pressure combustion completely contributes to the temperature rise [34]. With this approximation of the flame front temperature combined with the pressure inside the system, it is possible to calculate the species concentration of the mass flow that flows from the flame front into the burnt zone in a second step. The temperature of this mass flow is also stated as the adiabatic isobaric flame temperature.

With the approximation of the fuel combustion as chemical equilibrium, the question arises whether this simplification significantly affects the results. In order to investigate this issue, a steady-state flame is simulated with Cantera. A steady-state flame is defined as a 1D simulation of a freely propagating premixed laminar flame. Figure 3.9 shows the steady-state flame simulation result in terms of temperature (y-axis) and length (x-axis) and the corresponding CO mole fraction. The temperature of the unburnt pre-mixed gasoline surrogate and air mixture is 720 K, and the pressure is 30 bar. The dotted lines show the flame thickness, according to [26]. The fluid reaches the adiabatic isobaric flame temperature approximately 0.15 mm behind the flame front.

It can be observed that the CO mole fraction first increases and then decreases. This behavior is due to the breakup of long-chain hydrocarbons (CO increase) and the following oxidation of these newly formed CO (decrease). The final CO mole fraction of the flame is also reached approximately 0.15 mm behind the flame. In comparison with the volume profile of the burnt zone in Figure 3.10,

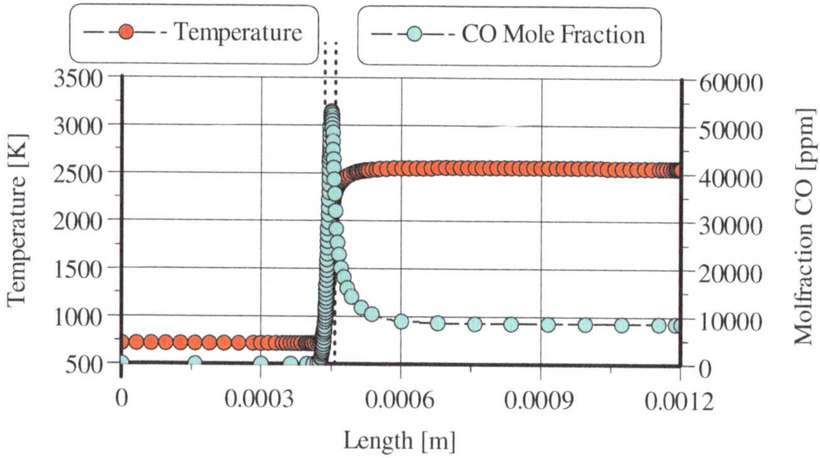

Figure 3.9: 1D steady-state flame simulation result: Temperature and *CO* mole fraction.

it can be said, that directly behind the flame and therefore directly with the entering into the burnt zone, the species mole fraction is approximately in chemical equilibrium. This fast establishing chemical equilibrium results from the high temperatures and the resulting high reaction rates close to the flame front. The expected deviation of chemical equilibrium and actual composition according to reaction kinetics occurs during the exhaust stroke. The temperatures are lower, and so are the reaction rates. Therefore, this simplification does not affect the results in a significant way. The results of a more complex model approach, including a reaction kinetic based flame model, would show similar results in terms of species concentration and mass flow temperature.

The change in species concentration inside the burnt zone is calculated by using the implemented reaction mechanism. The burnt zone is represented as a 0D reactor. For every time step, the temperature and pressure of the reactor are used to determine the current reaction rate of each defined reaction equation of the reaction mechanism.

The new species concentration inside the 0D reactor will result in a changed specific gas constant. In combination with the input data regarding volume and mass of the burnt zone, the temperature and pressure are determined. Hence, the system is not an isolated thermodynamic system; energy can pass the system's boundaries. This is necessary to model the wall heat losses of the burnt zone system. These also need to be implemented as input data as they need detailed data of the cylinder head and cylinder geometry (and others of course). The wall heat losses of the unburnt zone are not taken into account because they are already passively included in the unburnt zones temperature.

This procedure can now be applied to the 0D reactor for each time step, and because of this, the species concentration inside the burnt zone can be displayed from the moment that the burnt zone starts to exist until the exhaust valve open event. This approach does not need a threshold temperature to keep the species concentration constant.

An example engine operating point is chosen to give a detailed impression on the results in the following. The chosen engine operating point is carried out with the GT-POWER model presented in chapter 4. The engine operating parameters are listed in Table 3.3.

Table 3.3: Modelling of the burnt zone - engine operating point overview.

Engine Speed	Load	Valve Overlap	Average $\lambda_{Combustion}$
1600 rpm	180 Nm	90 °CA	0.85

Figure 3.10 shows the volume, and Figure 3.11 shows the mass of the burnt zone. These two time-dependent parameters result from the described two-zone combustion (chapter 2.3.2) and are necessary as input data for the reaction kinetics supported species determination. Both parameters are plotted from 5 °CA up to 181 °CA. This range covers the complete high-pressure phase of the engine cycle for one cylinder.

The course of the mass (Figure 3.11) gives an impression on the time span that is necessary for the flame front to reach the entire unburnt zone. The entire cylinder volume is declared as the burnt zone as soon as the corresponding mass does not increase anymore. Due to the moving piston, the volume of the burnt zone still increases.

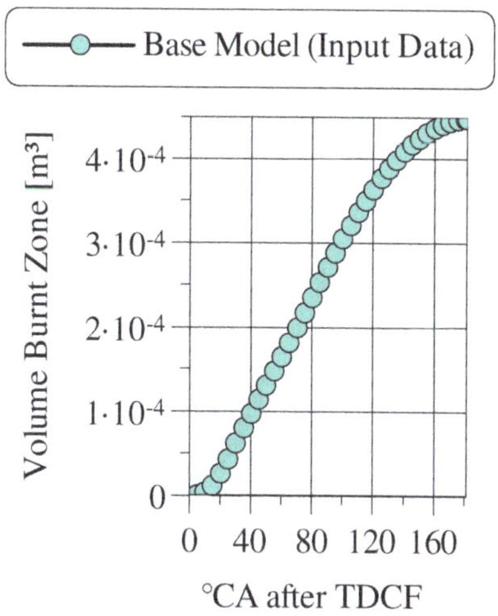

Figure 3.10: Volume of the burnt zone.

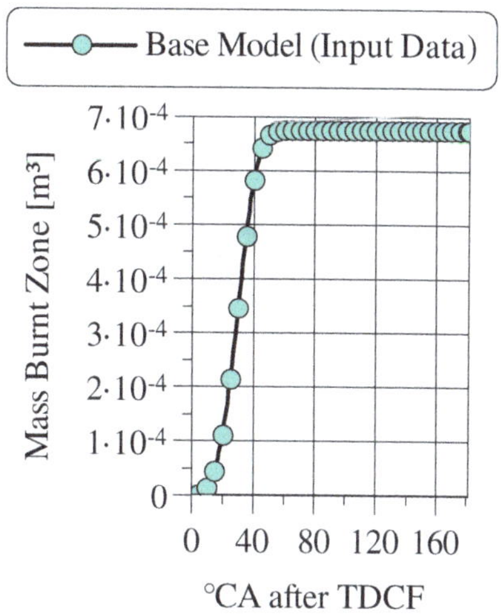

Figure 3.11: Mass of the burnt zone.

Figure 3.12 (a) and (b) display the course of temperature and pressure. The cyan colored markers do again show the input data as they did in Figure 3.10 and Figure 3.11. The blue markers show the results of the reaction kinetics calculation. As described above, the temperature and pressure both result from the change in volume, mass, species concentration and wall heat losses. The difference in pressure and temperature is due to the differences in species concentration.

Figure 3.13 (a) and (b) and Figure 3.14 (a) and (b) illustrate the mole fraction of selected species inside the burnt zone. The deviation in mole fraction during the time span with very high temperatures is due to different chemical equilibrium parameters (the literature values deviate slightly). Both cases are, because of the high temperatures, in or very close to the chemical equilibrium, which is slightly different.

The difference in the later stage of the high-pressure phase also includes the increasing deviation from chemical equilibrium to the actual occurring reactions. The lower temperature cause lower reaction rates. The deviation during the

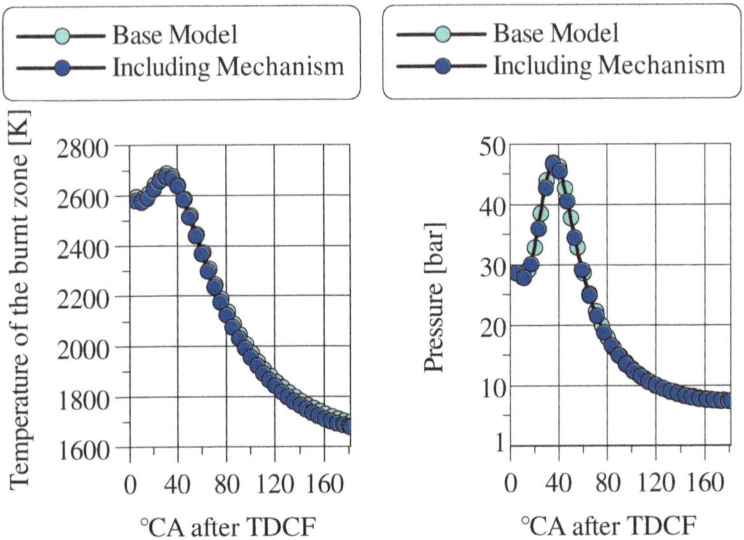

(a) Temperature (b) Pressure

Figure 3.12: Temperature and pressure of the burnt zone with the state of the art model as baseline case and the results of the reaction mechanism based calculation.

complete high-pressure phase is still quite small. The reasons are the high temperatures, and in case of the CO_2 and CO emissions, the freezing of the species concentration does compensate the difference due to the reaction speed reduction. In case of the H_2O and H_2 emissions, the freezing of the species concentration causes an increasing difference but with regard to the y-axis scaling, the difference is in a small order of magnitude. In addition to the findings of chapter 3.2.5, not the entire amount of CO and H_2 is oxidized during the post-oxidation process inside the exhaust manifold. Therefore, this deviation in raw emissions can be neglected during the post-oxidation model development process.

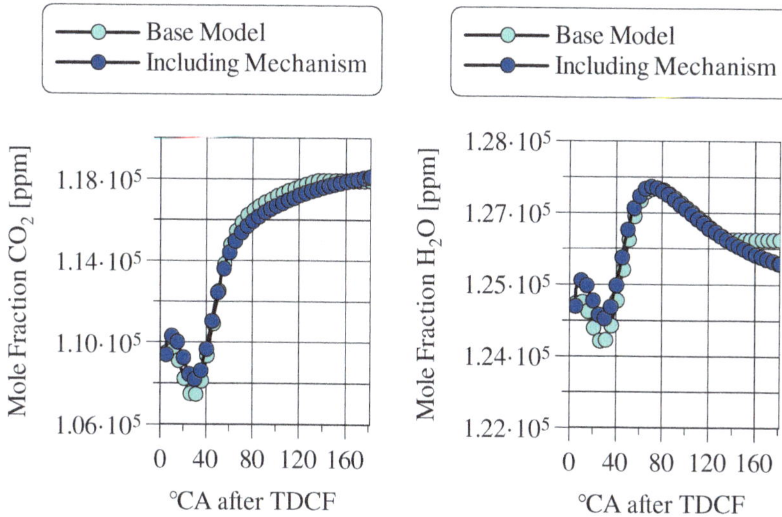

(a) CO_2 mole fraction

(b) H_2O mole fraction

Figure 3.13: Species mole fraction of the burnt zone with the state of the art model as baseline case and the results of the reaction mechanism based calculation.

This investigation can also be applied to the 1D engine that is the base of the post-oxidation model application discussed in chapter 5. The evaluation area is the low-end torque area due to its relevance in this study. In Figure 3.15 (a) and (b), the *CO* emissions' deviation is displayed as absolute deviation and relative deviation. The diagrams show the deviation of taking the time scale of chemical reactions into account. The maximum difference in the relevant area is approximately 1000 ppm of *CO*, and the maximum relative deviation is approximately 5 %. The highest relative deviation is close to 30 %, but the absolute deviation in this area is quite small (100 ppm – 500 ppm).

A similar evaluation is made with regard to H_2 emissions and displayed in Figure 3.16 (a) and (b). The maximum deviation is -1500 ppm, corresponding to a deviation of -10 %. Areas with a higher relative deviation show a small absolute deviation due to the low H_2 emissions.

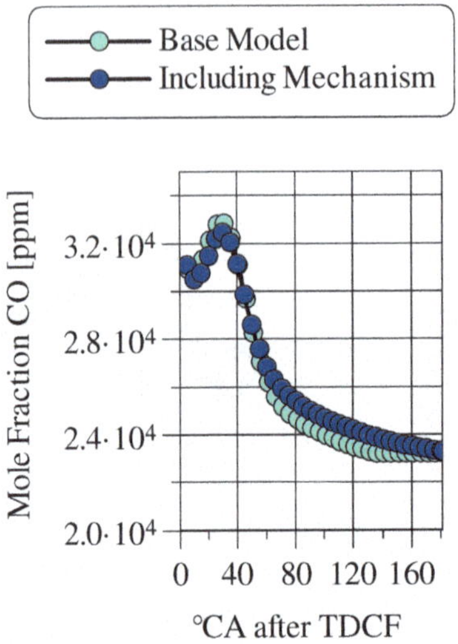

(a) CO mole fraction

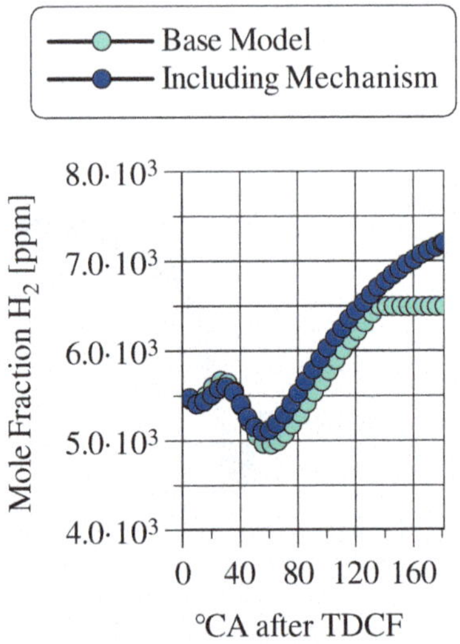

(b) H_2 mole fraction

Figure 3.14: Species mole fraction of the burnt zone with the state of the art model as baseline case and the results of the reaction mechanism based calculation.

This investigation shows that the simplification in emission calculation in terms of chemical equilibrium can be applied in this work. The maximum deviation is rather small, and the advantages in calculation time exceed the disadvantages of the emission calculation.

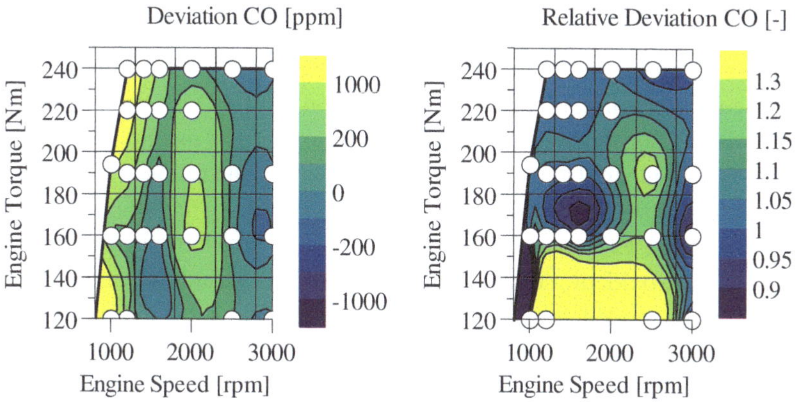

(a) Absolute Deviation (b) Relative Deviation

Figure 3.15: Absolute and relative *CO* deviation of the baseline case and the reaction kinetic based approach.

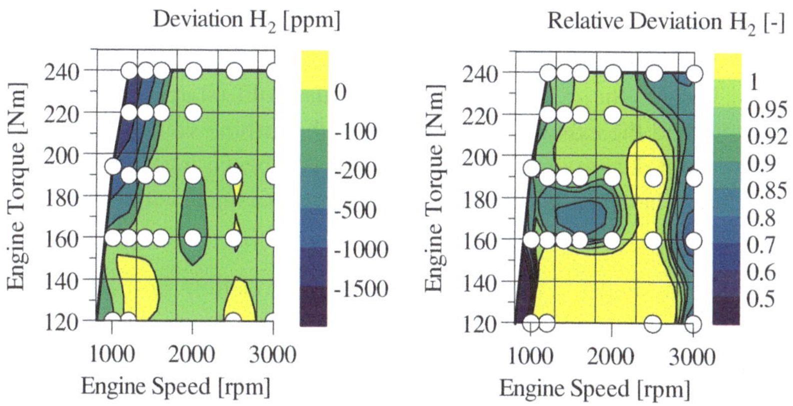

(a) Absolute Deviation (b) Relative Deviation

Figure 3.16: Absolute and relative H_2 deviation of the baseline case and the reaction kinetic based approach.

3.2 3D-CFD Simulation Results Evaluation

This chapter concentrates on the evaluation of the 3D-CFD results, which are presented and discussed in [1]. The shown 3D-CFD results during this chapter are – if not stated different – the results of the just mentioned 3D-CFD simulations. Compared to the evaluation made in [1], the following evaluation concentrates on the 1D model development and therefore, on different aspects. The evaluation bases on the question, which effects inside the exhaust manifold, affects the post-oxidation the most, how can these be quantified and how can these effects be modelled by the usage of a 1D engine model and its available tool-set.

3.2.1 Investigation Data Base

The subsequent investigation bases on the outcome of the 3D-CFD simulation described in [1]. The 3D-CFD simulation results are the outcome of two different types of 3D-CFD simulation. The first simulation type is a full engine simulation carried out with the simulation tool Quick-Sim [5].

Quick-Sim is a 3D-CFD tool that is developed to simulate internal combustion engines with a low dependency on initial conditions. The simulation domain includes, besides the cylinders, the intake manifold (starting right behind the throttle) and the exhaust manifold (up to the turbine volute intake). The necessary input data are pressure signals of a test-bench investigation of the same engine. These test-bench investigations are also made within the project, described in [1].

The full engine model is the result of a precise modelling of the test-bench engine. The model is very well tuned and able to display the combustion process of the test-bench engine with high accuracy.

The full engine simulation is able to reproduce the test-benchs indicated mean effective pressure with a difference of only 2 % and the maximum pressure inside the cylinder during the combustion with a difference of only 0.9 % - 4 %. The total air consumption including scavenging air of the validation engine operating points differs by 1 % to 4 %. More details on the full engine 3D-CFD simulation validation can be found in [1] and [33].

The validation of the full engine simulation results show that the simulation approach is able to reproduce the combustion process as well as the scavenging process of the test-bench engine with high accuracy.

The full engine model simulation is needed to produce boundary conditions at the position of the exhaust valve. These boundary conditions are used to run a second 3D-CFD simulation of only the exhaust manifold from the exhaust valves to the turbine volute.

This second 3D-CFD simulation is performed using the commercial 3D-CFD tool Star-CCM+. The simulation of the exhaust manifold only allows a finer discretization of the volume and the implementation of a reaction mechanism. The reaction mechanism permits the investigation of the post-oxidation inside the exhaust manifold in great detail. The used reaction mechanism inside the 3D-CFD mesh is reaction mechanism 2 [12], [27].

The validation of the fine mesh 3D-CFD simulation including a reaction mechanism is also documented in [1]. Here, beside the species at different positions, the temperature of the exhaust gas is evaluated and compared with the corresponding test-bench investigations.

The validation shows, that the temperature of the exhaust gas do match the results of the test-bench investigation. The validation is made at different positions along the exhaust manifold. It is important to stress, that the evaluation of the exhaust gas temperature is error prone. The reason for this is the radial temperature distribution inside the exhaust manifold and the small area of evaluation using a temperature measurement device (e.g. a thermocouple). Therefore, the direct comparison of the local temperature is less meaningful as the comparison of the temperature curve along the exhaust manifold. The 3D-CFD simulation results show a similar tendency in temperature increase due to post-oxidation effects.

The 3D-CFD simulations that are the base of this work are considered to be validated with test-bench measurement data. 3D-CFD simulation results offer a wider range of evaluable results in comparison to test-bench investigations. Especially the analysis of the chemical heat release due to post-oxidation and the location of this very chemical heat release make the usage of 3D-CFD simulation results inevitable. The requirements for the in this work newly developed 1D post-oxidation model is to display the chemical heat release due to post-oxidation and its location inside the exhaust manifold. Therefore, the newly developed 1D post-oxidation model bases on the the validated 3D-CFD simulation data (as provided by [1]) and is also validated by 3D-CFD simulation data. A validation with test-bench data would not be expedient here.

3.2.2 Available Engine Operating Points

The engine operating points are all chosen to have a valve overlap of 60 °CA. This overlap enables the usage of scavenging. Fresh air travels through the intake valve directly to the exhaust valve without participating in any combustion inside the cylinder. Hence, the TWC needs to be operated with nearly stoichiometric condition, the overall engine air-fuel ratio needs to be stoichiometric. This leads to the fact that the combustion inside the cylinders needs to be done with a rich composition. The level of enrichment inside the cylinders is determined by the amount of scavenging air passing the cylinder during the valve overlap. Therefore, the air-fuel ratio inside the cylinders cannot be chosen directly but indirectly by the duration of the valve overlap. The available engine operating points are listed in Table 3.4.

The available engine operating points are all located close to the low-end torque engine speed with a variation on the engine load. The chosen reaction mechanism is reaction mechanism 2 (compare Table 3.2) due to its comparatively low computation time. The chosen in-cylinder air-fuel ratio is a result of the overall stoichiometric condition and the valve overlap.

Two of these engine operating points are also simulated without implemented reaction mechanism to rate the mixing conditions of incomplete burnt exhaust gas and fresh scavenging air inside the exhaust manifold. This is discussed in more detail in chapter 3.2.6.

Table 3.4: Overview of the available engine operating points (3D-CFD simulation results).

Name	Engine Speed	Load	Valve Overlap	$\lambda_{Post-Ox.}$	$\lambda_{Comb.}$	Reaction Mechanism
Case #1	1200 rpm	150 Nm	60 °CA	1	0.96	2
Case #2	1200 rpm	174 Nm	60 °CA	1	0.89	2
Case #2a	1200 rpm	174 Nm	60 °CA	1	0.89	-
Case #3	2000 rpm	180 Nm	60 °CA	1	0.96	2
Case #4	1600 rpm	180 Nm	60 °CA	1	0.91	2
Case #4a	1600 rpm	180 Nm	60 °CA	1	0.91	-

In addition to the engine operating points listed in Table 3.4, case #4 is again simulated, including boundary condition changes. The target of these changes is a sensitivity analysis and a 1D post-oxidation model validation (described and discussed in chapter 4). Table 3.5 gives an overview of these engine operating points and each of their boundary condition changes.

Table 3.5: Case #4 based sensitivity analysis engine operating points (3D-CFD simulation results).

Name	Modification
Case #4b1	A decrease in the exhaust gas peak temperature by – 100 K: $\Delta T_{Ex.Gas}$ = -100 K
Case #4b2	$\Delta T_{Ex.Gas}$ = -300 K
Case #4c	Exhaust manifold wall temperature = 800 K: ΔT_{Wall} = 800 K
Case #4c1	ΔT_{Wall} = 800 K and $\Delta T_{Ex.Gas}$ = -100 K
Case #4c2	ΔT_{Wall} = 800 K and $\Delta T_{Ex.Gas}$ = -300 K

The decrease in emissions peak temperature is made for an exhaust temperature above 750 K to exclude unrealistic low emission temperatures. This means, if the temperature of the emissions is greater than 750 K, the temperature is decreased by the corresponding temperature correction. The minimum decreased temperature is 750 K. Temperatures below 750 K are not modified.

All engine operating points in Table 3.4 are carried out with adiabatic exhaust manifold walls. This is made to achieve increased comparability of 3D-CFD simulation and 0D/1D simulation results. Engine operating points with a stated exhaust manifold wall temperature do not have an adiabatic wall condition but a constant wall temperature. The exhaust manifolds wall temperature is set to 800 K. This constant value approximates realistic wall temperatures. This fixed temperature is chosen to maintain a more realistic environment for the exhaust gases while being applicable for both simulation environments: 3D-CFD and 0D/1D simulation.

3.2.3 Cylinder Emissions

The differences in combustion air-fuel ratio of each engine operating point lead a different amount of incomplete combusted emissions like CO and H_2. Figure 3.17 (a) and (b) show the differences in CO and H_2 emissions for each engine operating point. Both kinds of emission are plotted as the accumulated mass of all four cylinders during one engine cycle. The amount depends, beside of the combustion air-fuel ratio, on the absolute mass of injected fuel and therefore on the engine load. The calculation of these combustion products is made as described in [5].

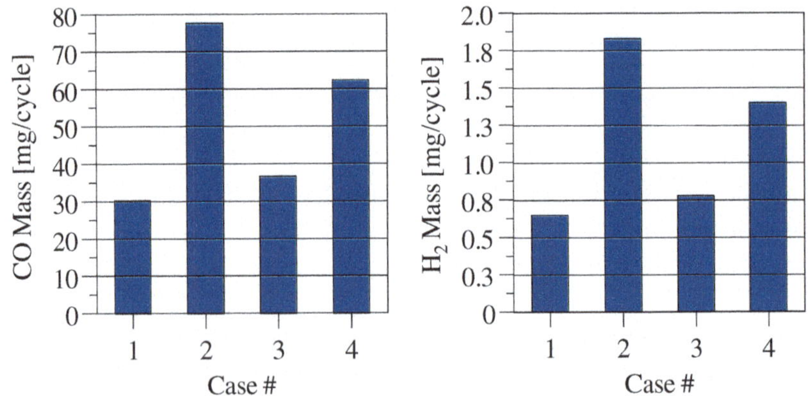

(a) CO emissions

(b) H_2 emissions

Figure 3.17: Emissions of all four cylinders during one entire engine cycle.

While the mass of *CO* emissions does exceed the H_2 emissions by roughly 45 times, the amount of releasable energy provided by each species does not differ in a completely different order of magnitude. Figure 3.18 displays this amount of energy emitted by all four cylinders combined in each cycle. This shows the importance of the realistic modelling of H_2 emissions. The H_2 constitutes about a quarter of the releasable heat of the incomplete combusted fuel.

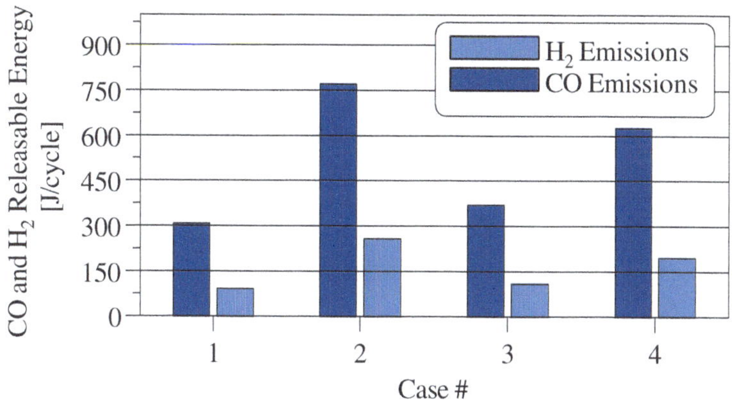

Figure 3.18: Accumulated releasable heat out of *CO* and H_2 entering the exhaust manifold during one entire engine cycle (all four cylinders) – based on lower heating value.

3.2.4 Evaluation of Heat Release due to Post-Oxidation Effects

This chapter is dedicated to the evaluation of the post-oxidation heat release inside the exhaust manifold. Different engine operating points are compared and discussed to identify how the differences in each case influence the post-oxidation. This includes a well-thought distribution of the 3D-CFD mesh in subvolumes and the definition of a quantifiable post-oxidation rate. At first, the evaluation subvolumes and the evaluation surface positions are defined. This is followed by the evaluation of the chemical heat release.

Definition of evaluation areas and surfaces

The 3D-CFD detailed mesh is divided into several subvolumes and several surfaces perpendicular to the flow direction. The subvolumes enable the spatial evaluation of chemical heat release, and the surfaces allow the evaluation of mass flow rates at different positions.

The subvolumes contain the valve area, which is located close to the exhaust valves. This part includes the boundary condition cells of the valve mass flows and the merging of the mass flows of the two exhaust valves of each cylinder. The second subvolume is the adaptor area. These four 10 cm adaptors are necessary for the test-bench measurement investigation to apply the measurement devices. The third subvolume is the mixing area, characterized by the junction of the four single pipe systems. As soon as the pipes are merged into a single pipe, the fourth area is called turbine area. A schematic illustration of the subvolumes is given in Figure 3.2.4.

Figure 3.19: Schematic overview of the evaluation subvolumes. Blue: valve area, green: adaptor area, yellow: mixing area, red: turbine area.

The evaluation surfaces are placed at each subvolume transition and within these subvolumes. The valve area is cut by ten surfaces starting from a surface directly at the mesh intake and ending at the transition into the adaptor area.

The adaptor area contains four evaluation surfaces, one at the transition from the valve area, one at the transition to the mixing area and two additional surfaces with constant spacing in between of those two. The mixing area and the turbine area do also have evaluation surfaces at the area transition points and the transition into the turbine volute. The evaluation is carried out between the boundary conditions at the entrance of the valve and the surface at the transition from the turbine area to the turbine volute. This evaluation area is defined in this way in order to avoid numerical issues at the turbine volute. An overview of the volumes of each subvolume is given in Table 3.6.

Table 3.6: Overview of the proportions of each subvolume.

Subvolume	Volume [l]
Valve area	0.39
Adaptor area	0.31
Mixing area	0.45
Turbine area	0.08

Evaluation of Released Heat

Figure 3.20 illustrates the chemical heat release inside the exhaust manifold. The exhaust manifold heat release is a result of the accumulation of the heat release of each subvolume mentioned above. The released heat is quite proportional to incoming chemically bound heat (compare Figure 3.18).

The relation between the incoming amount of releasable heat and the actually released heat inside the exhaust manifold is used in the following as post-oxidation rate, as stated in equation 3.13:

$$\text{Post-oxidation rate} = \frac{\text{Chemical heat release}}{\text{Incoming releasable heat}} \tag{3.13}$$

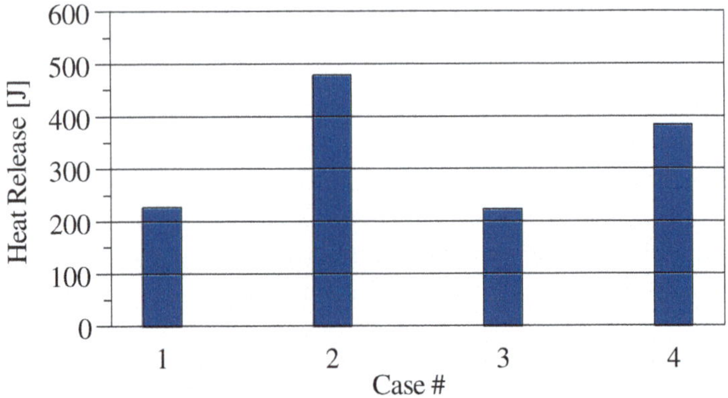

Figure 3.20: Heat release inside the evaluated exhaust manifold areas during one entire engine cycle due to post-oxidation.

This definition allows the comparison of the different engine operating points with each different amount of incoming emissions. The post-oxidation rate of case #1, #2, #3 and #4 is displayed in Figure 3.21. The resulting post-oxidation rate of case #2 up to #4 is quite similar (approximately 46 %). Since the boundary conditions of these cases do still differ, the post-oxidation process inside the exhaust manifold seems to be a complex combination of different influencing parameters. The post-oxidation rate of case #1 is the highest with approximately 57 %.

The heat release can be evaluated in each defined subvolume according to the classification made above. The corresponding subvolume heat release for each engine operating points is shown in Figure 3.22. It can be clearly observed that the heat release mainly takes place in the mixing area of the exhaust manifold.

Hence, this area is the greatest in terms of volume (compare Table 3.6), the evaluation needs to be done with a volume-specific heat release to erase this influence. Figure 3.23 shows the volume-specific heat release of each subvolume. This illustration shows that the later the emissions reach the subvolume, the higher the volume-specific heat release. This point of view also needs to be interpreted with caution, since the turbine area is the only area, which needs to be passed completely by all four cylinders emissions, while

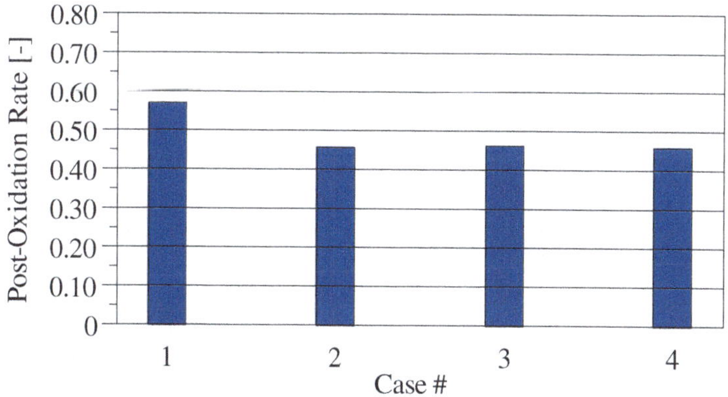

Figure 3.21: Post-oxidation rate of each case during one entire engine cycle.

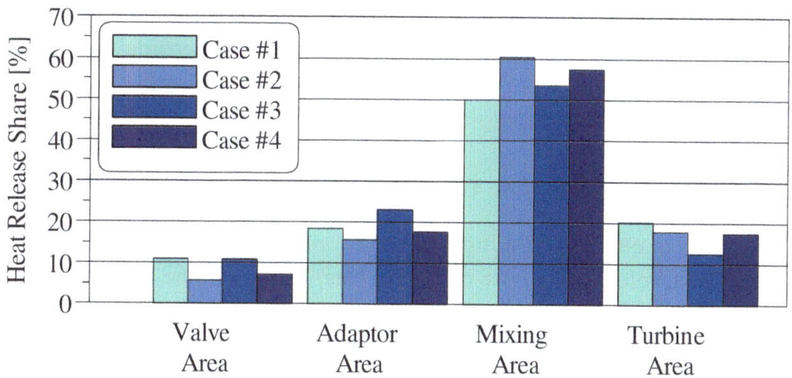

Figure 3.22: Heat release distribution inside the exhaust manifold of each simulated 3D-CFD simulation.

the bulk of the mixing area is passed by a maximum of two exhaust gas mass flows. This lowers the volume-specific heat release. Still, the diagram shows that the post-oxidation does not appear directly after the exhaust gas flow into the exhaust manifold. In between the exhaust valves and the turbine volute, a complex interaction of mixing, chemical ignition delay and self-reinforcing takes place. This interaction of influencing parameters is investigated and discussed in the following.

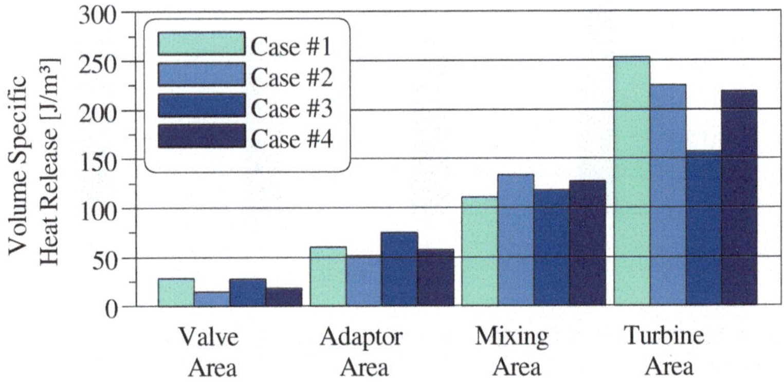

Figure 3.23: Volume specific heat release distribution inside the exhaust manifold of each simulated 3D-CFD simulation.

Sensitivity Analysis Based on Case #4

Subsequent, the engine operating points, according to Table 3.5, are evaluated regarding heat release and post-oxidation rate. These engine operating points are based on case #4 but with boundary condition changes by means of exhaust gas peak temperature and wall boundary condition variations. The resulting post-oxidation rate inside the exhaust manifold mesh is displayed in Figure 3.24. Case #4 is again plotted as a baseline case.

With decreasing exhaust gas temperature, the chemical heat release decreases by 17 % with an exhaust gas peak temperature decrease of 100 K. The decrease of 300 K leads to an 80 % decrease in heat release. A change of -100 K does not significantly impact the heat release inside the exhaust manifold because the temperature increases during the post-oxidation process. As long as the temperature is high enough to start the post-oxidation, a self-sustaining effect begins. This effect bases on the temperature increase of mixed emission clouds (spatially limited clouds of mixed exhaust gas and scavenging air), whose temperature is too low to start the oxidation effect. If the critical amount of heat release is not achieved inside the exhaust manifold to heat these mixed clouds, the self-sustaining effect cannot start.

The change of the wall boundary condition from adiabatic to a constant wall temperature of 800 K leads to a decrease in heat release of approximately 9 %. An additional decrease of the exhaust gas peak temperature as done before enhances the decrease up to 30 % (for a temperature decrease of 100 K) and 81 % (temperature decrease of 300 K). The wall heat temperature of 800 K enables the heat flux through the exhaust manifold walls. This heat flux decreases the temperature inside the exhaust manifold, and therefore the reaction rate of the post-oxidation process decreases. The change of the exhaust gas peak temperature is a more direct way to achieve the same effect.

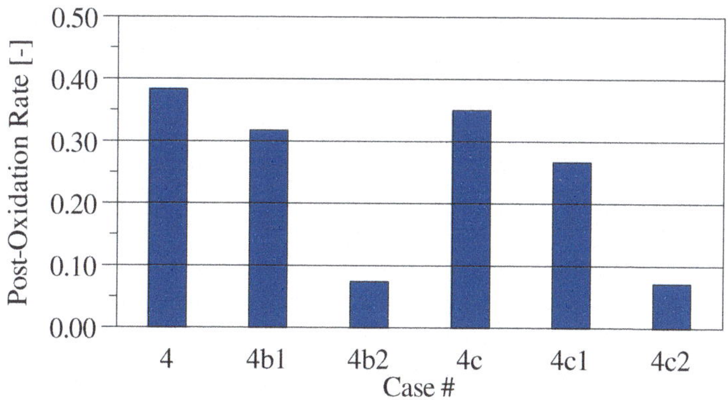

Figure 3.24: Post-oxidation rate of a sensitivity analysis, based on case #4.

Figure 3.26 shows the released heat of case #4, case #4b1 and case #4b2 inside each subvolume. The by 100 K reduced peak temperature results in a heat release in the valve area that is more than halved compared to the baseline case (case #4). Since the valve area has a quite low volume, the residence time of the exhaust gas is also quite low. In combination with a low temperature and the accompanying long chemical ignition delay, this leads to a decrease of heat release in this area. However, some of the exhaust gases are mixed, and the temperature is high enough to start the post-oxidation with a very short ignition delay.

The heat release inside the valve area increases the temperature of the mixed exhaust gas clouds travelling through the following areas. Since the mixing of the exhaust gas is not affected by the exhaust gas peak temperature, it can be assumed, that the same amount of exhaust gas and air are mixed up compared to the baseline case. The lower heat release delays the post-oxidation process, but as soon as a certain amount of heat is released, the self-reinforcing effect starts. The self-reinforcing effect also needs time to establish. Therefore, this effect is stronger, the longer the covered distance. Close to the turbine, the difference is only 8 %.

In Figure 3.25, the heat release share inside the exhaust manifold is plotted. According to the findings so far, the heat release distribution is relocated to areas more downstream in comparison to the baseline case. The lower temperature starts the self-reinforcing effect with a delay.

If the peak temperature is lowered by 300 K, the amount of mixed exhaust gas clouds, whose temperature is high enough to start a post-oxidation process with a short ignition delay is again lower. The valve area's heat release is again approximately 40 % lower than in case #4b1 (Figure 3.26). The main difference in these cases is that in case #4b2, the amount of heat release cannot increase the temperature to start the self-reinforcing process within the exhaust manifold. This can be observed in Figure 3.26. The increase of heat release in the adaptor area compared to the valve area is not as big as for the baseline case and case #4b1.

The heat release share in the mixing area is similar to case #4 and #4b1 (Figure 3.25), while the heat release share close to the turbine is less than 50 % of the baseline case. Based on the results of case #4b1, the heat release share should be higher as the heat release share of the baseline case as well as case #4b1 due to the delayed self-reinforcing effect. This means that the self-reinforcing effect should increase further, but instead, the self-reinforcing effect weakens.

This opposite behavior results from the high mixing inside the mixing zone (discussed in detail in chapter 3.2.6) and the associated decrease in temperature of the mixed clouds. This decrease in temperature happens because of the fact that the heat release occurs in regions, which combine incomplete burnt emissions, unburnt scavenging air and high temperatures. The exhaust gas peak temperature decrease causes that the necessary temperature is mainly reached because of the heat release inside the exhaust manifold. The low amount of heat release suggests the conclusion that areas that combine all three requirements are limited. Since these clouds host the fluid with the highest temperature, a further mixing with the surrounding gas lowers the temperature of the clouds. This leads to longer chemical ignition delays, limited heat release and the still ongoing mixing lowers the mixed exhaust gas clouds temperature even more. In case #4 and #4b1, the temperature decrease due to the mixing is compensated by the already high temperature and the still ongoing heat release. As soon as the heat release cannot compensate the decrease in temperature due to the mixing, the mixing in the exhaust manifold has a flame extinguishing effect.

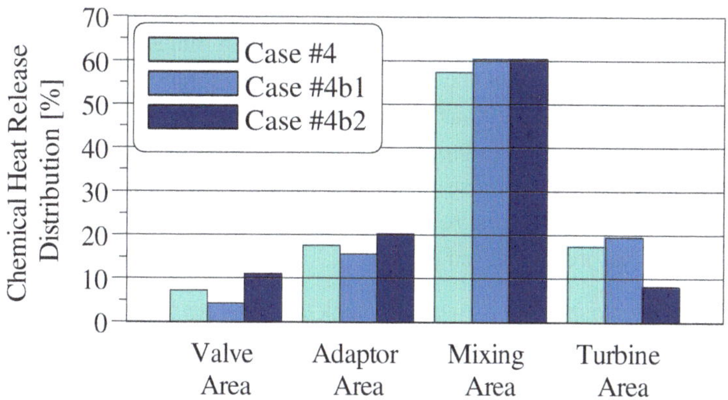

Figure 3.25: Spatial distribution of the chemical heat release share inside the exhaust manifold with decreasing exhaust gas peak temperature.

The next step contains an additional change of the boundary condition by means of a constant wall temperature of 800 K (case #4c). Then, the exhaust gas peak temperature is again adjusted according to Table 3.5 (case #4c1 and case #4c2).

The adjustment of the wall boundary condition (case #4c) causes an increase of the heat release close to the exhaust valve. This might be since the exhaust

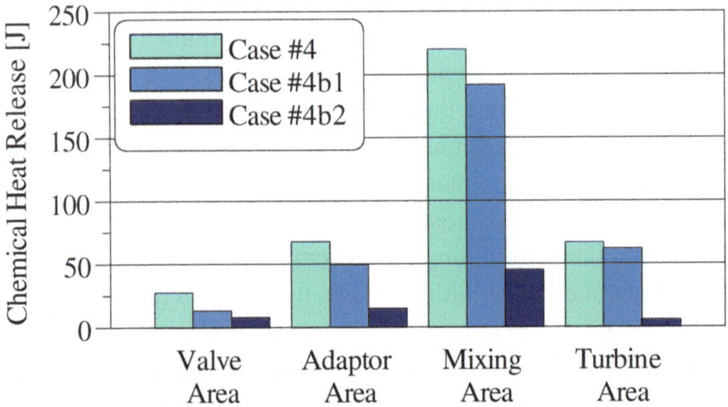

Figure 3.26: Spatial distribution of the chemical heat release inside the exhaust manifold with decreasing exhaust gas peak temperature.

manifold is heating the accumulated scavenging air in this area (discussed in chapter 3.2.5). This heating of the scavenging air counteracts the temperature decrease due to mixing effects, as discussed above and shortens the ignition delay. In the following areas (adaptor area, mixing area), the heat release falls behind the baseline case because of the heat losses at the exhaust manifold walls. The higher temperatures, resulting from post-oxidation, cause a higher heat flux and, therefore, increased cooling.

The additional decrease of the exhaust gas peak temperature by 100 K (case #4c1) compensates the heating effect of the wall in the valve area. As a result, the released heat is similar to the baseline case. The similar amount of heat release in the adaptor area (compared to case #4c) can result from the shorter ignition delay in case #4c. A larger proportion of the exhaust gas mixed directly after the exhaust valve is already oxidized in the valve area. In case #4c1, the longer ignition delay causes a heat release of this early mixed gas more likely in the adaptor area instead of the valve area. A combination of this longer ignition delay and a lower exhaust gas temperature leads to lower heat release in the valve area and a similar heat release in the adaptor area compared to case #4c. However, the increase in temperature is not as high as in case #4c, and as a result, the released heat in the mixing and turbine area is lower.

Case #4c2 shows a similar behavior as case #4b2 (same temperature adjustment but adiabatic walls (Figure 3.26)). Although the share in the valve and the adaptor area is greater than for the case with adiabatic walls (most likely because of the heating of the scavenging air), the released heat is not enough to establish a self-reinforced post-oxidation process within the exhaust manifold.

Summarizing, the decrease of the exhaust gas peak temperature leads to a decrease in the resulting post-oxidation heat release. This is because of the lower reaction rates, caused by the lower temperature, leading to longer ignition delays and results in a delay of the self-reinforcing effect. If the temperature decrease is too high, the delay of the self-reinforcing effect gets too high. In combination with high mixing effects, even a flame extinguishing effect can slow down the post-oxidation additionally and prevent the self-reinforcement from establishing.

The change of the exhaust manifold wall boundary condition from adiabatic to a constant value of 800 K can shift the heat release share to happen in a greater proportion at the valve area and in a less high share in the mixing area.

This investigation shows that the post oxidation process inside the exhaust manifold is a complex interaction of mixing and reaction kinetics effects. Without mixing, the post-oxidation process cannot start, regardless of the fluid temperature. If the mixing is too high compared to the heat release, the temperature can decrease, and the post-oxidation effect is limited or even extinguished. In the following, the mixing of the incomplete burnt exhaust gas with the scavenging air is investigated in detail.

3.2.5 Evaluation of Emission Oxidation due to Post-Oxidation Effects

The investigation on the heat release inside the exhaust manifold raises some open questions on the behavior of the emissions inside the exhaust manifold.

The investigation on the volume specific heat release leads to the assumption that the reaction rate of the post-oxidation process is the reason for the heat release distribution inside the exhaust manifold. The chemical ignition delay and the consequent low heat release rate for areas close to the exhaust valves are followed by a self-reinforcing oxidation process. The more OH-radicals are

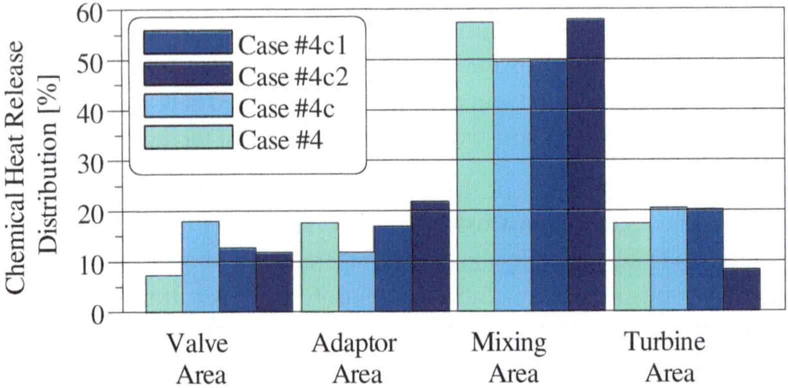

Figure 3.27: Spatial distribution of the chemical heat release share of the 3D-CFD exhaust manifold simulation results.

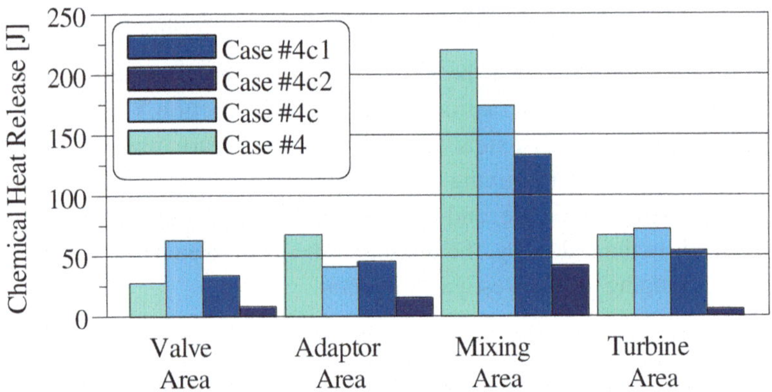

Figure 3.28: Spatial distribution of the chemical heat release of the 3D-CFD exhaust manifold simulation results.

produced, the more CO and H_2 can be oxidized, and the increased heat release leads to an increased temperature. This again increases the production of OH. This is the reinforcing effect of the post-oxidation process. However, a closer look at the behavior of the emissions inside the exhaust manifold reveals a second influencing parameter. Figure 3.29 shows the mass fraction distribution of O_2 inside the exhaust manifold during the exhaust stroke of cylinder number four (on the right-hand side). While the red areas highlight the areas with very high amounts of O_2, do the blue areas show the absence of this O_2. These areas highlight the incomplete burnt combustion products out of cylinder number four.

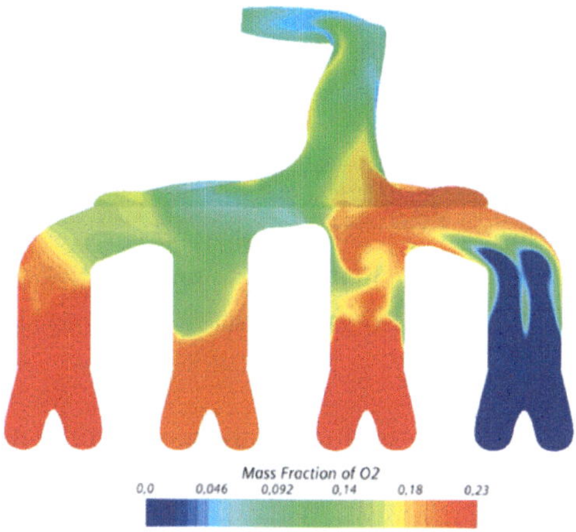

Figure 3.29: 2D cut of the 3D-CFD simulation results illustrating the O_2 mass fraction inside the exhaust manifold [1].

The incomplete burnt combustion products are flowing behind the O_2 through the exhaust manifold. The two clouds interact in a quite small area by mixing up. This mixing is indicated by a decreasing amount of O_2 mass fraction (green areas). The corresponding heat release rate of this specific time step is shown in Figure 3.30.

Figure 3.30: 2D cut of the 3D-CFD simulation results illustrating the chemical heat release rate inside the exhaust manifold [1].

It can be clearly observed that the areas with the most relevant heat release are located in the mixing areas of O_2 and incomplete burnt combustion products. The area of heat release is also very limited and characterized by a sharp demarcation. This shows a high dependency of the post-oxidation effect from the mixing inside the exhaust manifold.

The already mentioned high mixing rate in the mixing area is, according to the image in Figure 3.29, mainly enhanced by a vortex system. This vortex system appears because of the increasing pipe cross-section and the detachment phenomena occurring on edges, whose curvature is too high for the flow to follow. The vortex system prevents the scavenging air from flowing directly into the turbine, and the following exhaust gas is able to catch up and is also affected by the vortices. The vortex system also influences the mixing inside the turbine area, and therefore, the mixing in this part of the exhaust manifold is also increased. On the other hand, the geometry of the adaptor with its constant pipe cross-section prevents the flow from establishing such vortices. The shape of the mixing zone with its two characteristic spikes origins in the junction of the two pipes very close to the exhaust valve, upstream of the adaptor. The stable shape of the mixing zone underlines the theory that the adaptor area does not affect the mixing of the flow in a significant way.

Beside this, a second important finding can be made in Figure 3.29. This illustration shows an accumulation of O_2 near to the closed exhaust valves. The scavenging air (symbolized by the O_2 mass fraction) does not flow directly into the turbine but is pushed back to the exhaust manifolds by the valve opening event of another cylinder. At this specific time step, the most right cylinder is preventing the scavenging air from flowing to the turbine. The high pressure inside the cylinders results in the displacement of the O_2, due to the relatively low pressure during the valve overlap. The incomplete burnt combustion products are then pushed into the residual scavenging air of the cycle before. The high velocity of the exhausted combustion products leads to mixing with the residual scavenging air mentioned above. The accumulation of the scavenging air close to the exhaust valve due to a cylinder-to-cylinder interaction leads to the thought that besides the exhaust manifold geometry also the engine operating point influences the mixing effects inside the exhaust manifold. A higher engine speed could lead to a different cylinder-to-cylinder interaction as well as a change in engine load, due to the changing combustion products velocities.

Evaluation of the Species Conversion

The subsequent chapter focuses on the change in species concentration inside the exhaust manifold. The evaluation is made at the transition positions of the mesh areas defined in chapter 3.2.4. The used surfaces are listed in Table 3.7 and an illustrated overview is given in Figure 3.31.

Table 3.7: Overview of the defined evaluation surfaces.

Surface Location	Surface Name
Boundary Conditions – Exhaust Valve Area	Position 1
Exhaust Valve Area – Adaptor Area	Position 2
Adaptor Area – Mixing Area	Position 3
Turbine Area – Outlet	Position 4

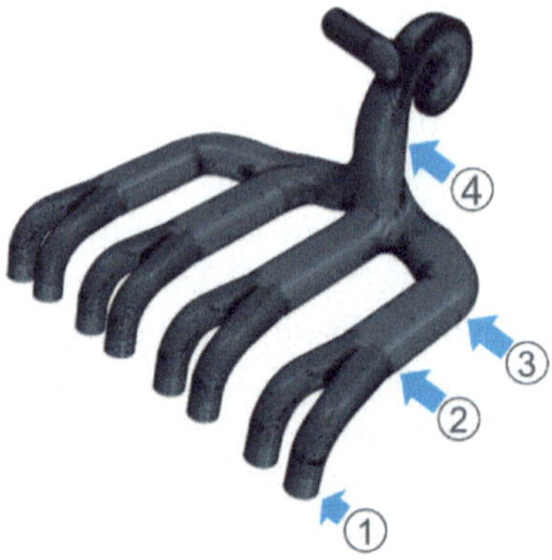

Figure 3.31: Overview of the defined evaluation positions according to Table 3.7.

The surfaces of position 1 up to position 3 exist for each cylinder individually. To keep track of things, each surface of position 1 up to position 3 is not evaluated individually, but the tracked species mass is summed up at each position. This represents a simplification, but it still suits the requirements of this investigation.

In Figure 3.32 (a), (b) and (c), the species mass of CO, H_2 and O_2 is tracked. The species mass is accumulated at the corresponding surface over one engine cycle and normalized by the boundary condition mass at the mesh inlet to achieve increased comparability.

The heat release distribution already showed that the highest amount of oxidized emissions needs to be between position 3 and 4. In terms of species mass, this theory can be confirmed. The CO and H_2 mass is decreasing by approximately 10 % to 20 % between positions 1 and 3, representing the valve and the adaptor area and decreasing by 30 % to 40 % between position 3 and position 4, representing the mixing and turbine area.

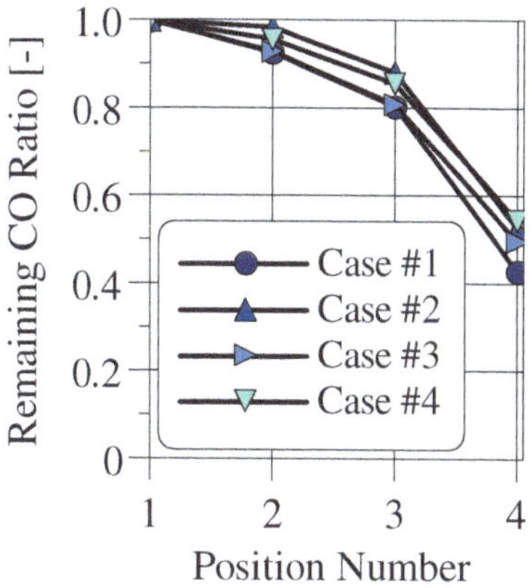

(a) Normalized accumulated CO mass

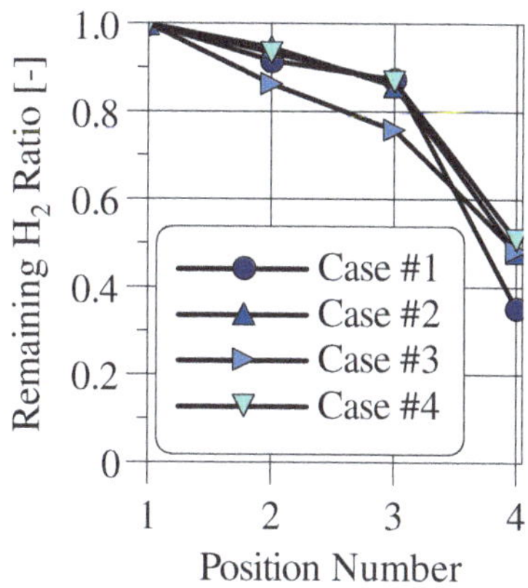

(b) Normalized accumulated H_2 mass

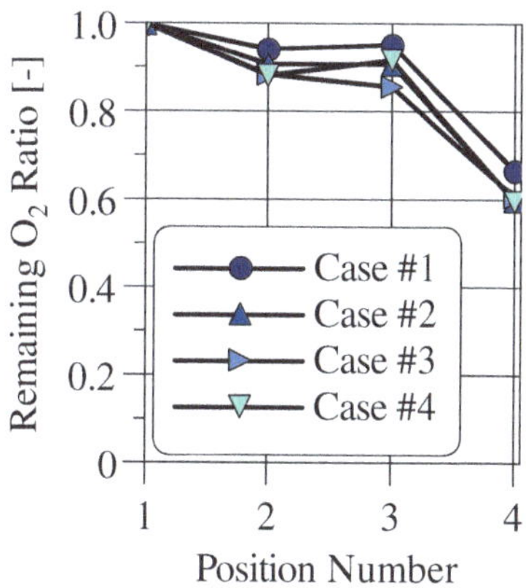

(c) Normalized accumulated O_2 mass

Figure 3.32: Normalized accumulated species mass at different positions during one entire engine cycle.

The percentage decrease of H_2 between position 1 and 2 is higher as the CO decrease. This happens because the H_2 oxidation does not necessarily need the OH-radical to oxidize (compare chapter 3.1.2). With increasing OH-radical formation, the H_2 mass percentage decreases. The decrease between position

1 – 2 and position 2 – 3 is roughly linear. The decrease of CO mass percentage, on the other hand, is at first quite restrained but increases between position 2 – 3 due to the increasing amount of OH.

The decrease of the O_2 mass percentage between position 1 and 2 can be associated with the production of OH-radicals. The limited decrease of O_2 between position 2 and 3 could result from the mixing situation inside the exhaust manifold. As mentioned before, the adaptor area does not show the geometrical requirements to develop a high flow mixing effect. This can now also be observed since the O_2, which flows in between of position 2 and 3 does not seem to be mixed up with a reaction partner. The decrease of CO and H_2 in this area happens to the already produced OH in their immediate surroundings. In between of position 3 and 4, the decrease of O_2 starts again and is, most likely because of the high mixing level, faster than in between of position 1 and 2.

The tracking of specific species masses inside the exhaust manifold underlines that the post-oxidation process is influenced by the mixing of the exhaust gas and the scavenging air on the one hand but also on the reaction rate of these mixed species on the other hand. The reaction rate again is also influenced by the mixing effects by means of decreasing temperature, as mentioned in chapter 3.2.4.

3.2.6 Development of a Post-Oxidation Evaluation Methodology

The post-oxidation heat release inside the exhaust manifold shows that the distribution of this very heat release is not proportional to the size of the corresponding subvolume. The reaction kinetics investigations, made in chapter 3.1, imply a relevant influence of the reaction rate and chemical ignition delay on the heat release location inside the exhaust manifold.

The more detailed view on the emission distribution inside the exhaust manifold and the formation of emission-air clouds indicates that also the mixing behavior inside the exhaust manifold is a size to be reckoned with.

The following subchapter is dedicated to the question, whether the reaction kinetics or the mixing effects inside the exhaust manifold are the most limiting factors for the post-oxidation and what effect should be focused for the 1D post-oxidation model development.

The evaluation of mixing as well as reaction rate influences make a quantification necessary, which is able to illustrate the importance of each. In the following, this quantification methodology is developed and applied to the 3D-CFD results.

Analysis Methodology

The analysis of the limiting factors of the post-oxidation process needs an appropriate tool for quantifying these factors.

The 3D-CFD simulation allows the evaluation of each defined cell on its own at any time step. Besides the local temperature, pressure and mass, the local species composition can be analyzed. This enables the investigation of local mixing conditions inside the exhaust manifold at any time step for each cell on its own.

The mixing condition needs to be quantified in order to be analyzed. The quantification of a mixing composition in the field of engine development is usually made by usage of the air-fuel ratio. Since the overall air-fuel ratio of the engine is stoichiometric, a stoichiometric air-fuel ratio inside the exhaust manifold would indicate a perfectly homogeneously mixed flow. As soon as mixing is not perfectly homogeneous, the air-fuel ratio is locally rich and locally lean – because of the overall stoichiometric condition. In locally rich mixtures, the post-oxidation cannot completely occur, since the necessary oxidizer is missing. In locally lean mixtures, the oxidizer will run out of reaction partners like CO and H_2.

The calculation of the air-fuel ratio is made according to equation 3.14.

$$\lambda = \frac{m_{air}}{L_{st} \cdot m_{fuel}} \qquad 3.14$$

The minimum required air depends on the used fuel and is calculated based on its $C_x/H_y/O_z$ ratio according to [14] by equation 3.15.

$$L_{st} = \frac{(x+0.25y-0.5z)137.644}{12x+y+16z} \qquad 3.15$$

Using the air-fuel ratio for mixture evaluation is not suitable. This should be demonstrated by the following thought experiment:

The definition of the air-fuel ratio is based on masses of each, fuel and air. This definition implies that the air-fuel ratio remains constant during the combustion and therefore the emissions of a combustion can be assigned an air-fuel ratio. A simplified rich combustion A produces H_2O, CO_2, CO and H_2. This combustion products of combustion A show the same air-fuel ratio as the reactants before the combustion. The second and even more rich combustion B produces similar products but with a higher ratio of CO and H_2 than combustion A.

After both example combustion processes are finished, and the reactants are converted into combustion products, the thought experiment goes on as follows. Since the combustion products do have by definition the same air-fuel ratio as the reactants (discussed above), the addition of O_2 to the combustion B products can increase the air-fuel ratio to match the air-fuel ratio of combustion A.

Now, both combustion products share the same air-fuel ratio, but the products of combustion B are mixed in a way that can oxidize some of the remaining incomplete combusted fuel (namely CO and H_2 in this simplified case) due to available O_2. The products of combustion A also contain CO and H_2 but no available O_2.

The amount of available O_2 can still be evaluated by consideration of the original air-fuel ratio of combustion A and B. Then an air-fuel ratio increase of the combustion products of combustion B would imply the availability of O_2. If the combustion products' air-fuel ratio is higher than the air-fuel ratio during the combustion, there are available oxidizers mixed to the combustion products.

However, the 3D-CFD evaluation of the exhaust gas inside the exhaust manifold does not allow identifying the original air-fuel ratio of the exhaust gas during its combustion. As stated in [1], the air-fuel ratio inside a cylinder differs locally from the mean value. This is because of the inhomogeneous mixture formation inside the exhaust manifold. As a result of this, the local air-fuel ratio of the exhaust gas inside the exhaust manifold cannot make a statement concerning its availability of oxidizers.

This theoretical consideration shows that the air-fuel ratio of a mixture is not suitable for a mixture evaluation. A rating of the oxidation potential of CO and H_2 needs to include information on the available oxidizer.

The evaluation of the local mixture inside the exhaust manifold cells needs a quantification method, which can display the theoretically possible amount of incomplete combusted fuel that could be oxidized by means of available oxidizer. As mentioned in chapter 3.1.2, the oxidation process of CO and H_2 is mainly driven by the radicals OH and HO_2, whose source again is O_2. Therefore, these three molecules are now declared as oxidizers. The main source of heat release is, in case of simulations according to [1], CO and H_2.

The local ratio of available oxidizer in relation to necessary oxidizer for complete combustion is defined as ι in equation 3.16.

$$\iota = \frac{2\chi_{O_2} + \chi_{OH} + 2\chi_{HO_2}}{\chi_{CO} + \chi_{H_2}} \tag{3.16}$$

$\iota > 1$ More oxidizer available than necessary
$\iota = 1$ Stoichiometric condition
$\iota < 1$ Lack of oxidizers

The evaluation of a mixture with the usage of ι is independent of the air-fuel ratio during the combustion. Unlike the air-fuel ratio, ι does distinguish between atomic oxygen, which is already burnt and partially or unburnt atomic oxygen.

For each time step and each cell, the local mass of CO and H_2 can now be evaluated. In combination with the local value of ι, a statement on the local mixing of incomplete combustion products and oxidizer can be made.

The mixing of combustion products and oxidizer is a spatially limited effect. Hence, the evaluation needs to be made for the finest possible available volume. In the case of 3D-CFD simulation, a single cell is the smallest possible discretization area for an evaluation. The evaluated cells are defined by surfaces, which cut the 3D-CFD mesh of the exhaust manifold perpendicular to the flow direction. The surfaces are chosen to be at defined positions inside the exhaust manifold (Table 3.7). All cells affected by the surface in this way are evaluated in the following individually and for each time step.

This methodology is applied to the surface, which marks the outlet of the evaluated part of the mesh. The incoming *CO* approaching this surface did not oxidize during its way through the exhaust manifold. This unoxidized *CO* can now be tracked for each cell and each time step to investigate the conditions of the *CO* at this point. This gives an impression, whether the *CO* did not oxidize because it was not mixed up with oxidizer or whether the *CO* had available oxidizer around, but the reaction kinetics was too slow to start the oxidation in time.

To visualize the evaluation, a histogram is chosen. During one whole engine cycle (720 °CA), each cell's CO mass is analyzed concerning ι. *CO* masses within the same range of ι are accumulated and visualized in a histogram. An example histogram is shown in Figure 3.33. The chosen engine operating point is case #4.

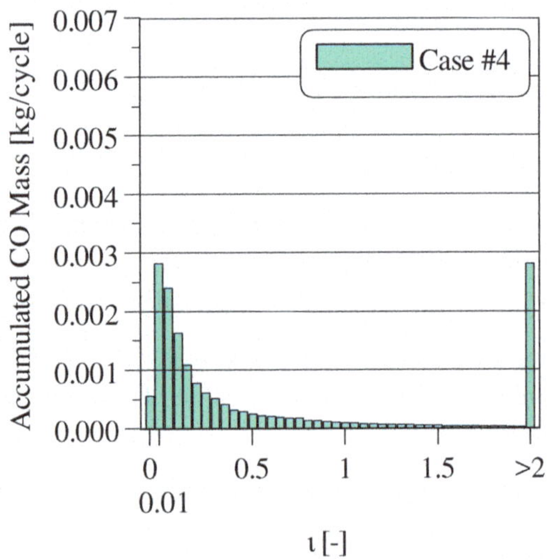

Figure 3.33: ι-histogram in position 4 – case #4.

To facilitate the interpretation of the histogram, Figure 3.34 shows the two achievable extremes of the evaluation as a schematic example illustration. The orange bars show the case in which the mixing inside the exhaust manifold is the limiting factor of the post-oxidation. This scenario shows an accumulation of remaining CO for ι close to zero and greater than two. If ι equals zero, then all CO approaching the mesh outlet does not have any oxidizer available. Since demixing is not likely to happen (second law of thermodynamics), all mixed CO must have been oxidized on its way through the turbine. The second orange bar for ι greater than two shows CO, which is badly mixed, because of the way too much fresh air. This great amount of fresh air does not only increase the amount of oxidizers but also the amount of N_2. With this great amount of inert gas, the thermal mass permits the increase of temperature due to heat release and therefore, this orange bar is also declared a mixing issue.

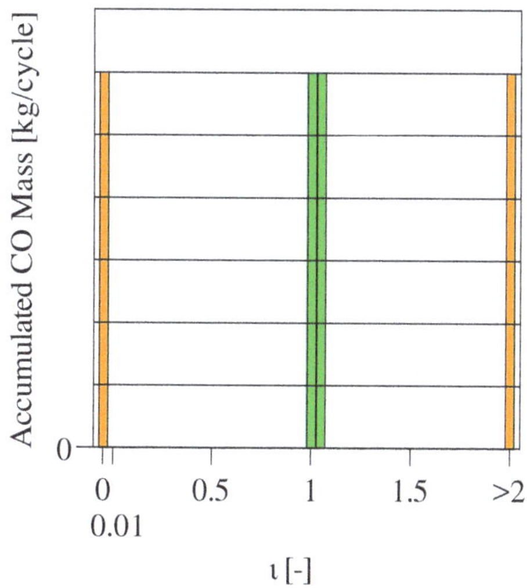

Figure 3.34: Schematic ι-histogram including two extreme case examples.

The green bars show the second achievable extreme with the reaction kinetics as the limiting factor. This example shows a perfectly mixed mass of CO approaching the evaluation surface position during one engine cycle. This implies that although the CO is perfectly mixed, it does not completely oxidize.

In this case, the limiting factor needs to be a too slow reaction rate for the time span available (travelling time from the exhaust manifold to the turbine volute).

A comparison of Figure 3.33 and Figure 3.34 indicate that the tendency of the CO mass condition at the mesh outlet shows a mixing dependent post-oxidation. The majority of the approaching CO is not oxidized until this point due to the lack of available oxidizer. However, there is still an amount of CO mixed up with oxidizer, which is not neglected.

The ι distribution of Figure 3.33 is located between the two extreme values illustrated in Figure 3.34. Figure 3.35 and Figure 3.36 show the ι distribution for case #2 and case #3. While the distribution in case #2 seems to be more mixing level depended, case #3 shows an opposite trend.

In order to quantify the ι distribution, the evaluation methodology needs to be additionally developed.

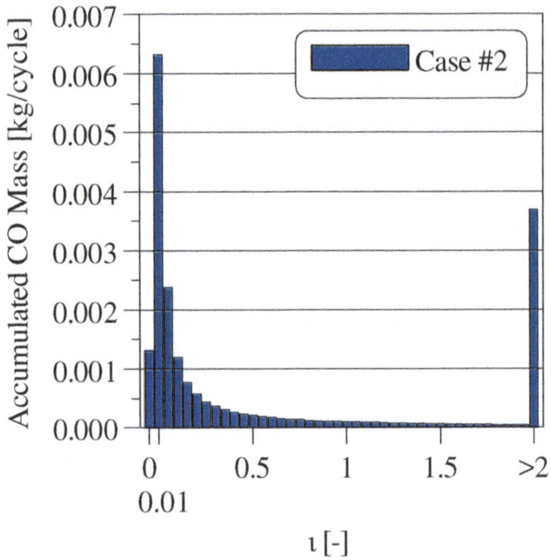

Figure 3.35: ι-histogram in position 4 – case #2.

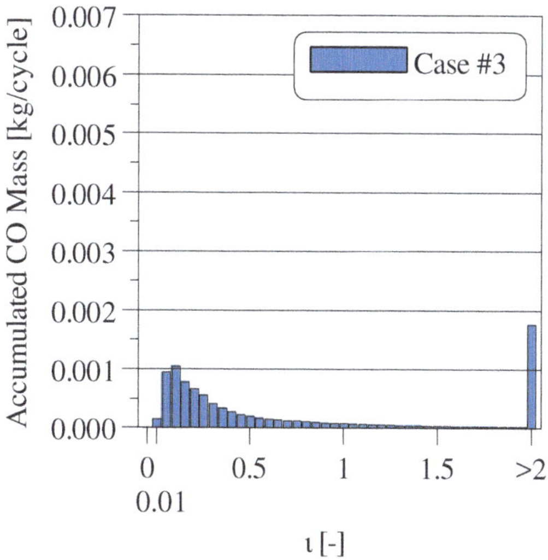

Figure 3.36: ι-histogram in position 4 – case #3.

The quantification of ι is based on the following assumptions and simplifications:

- A perfect mixing leads to a ι of 1.0.
- A decreasing mixing quality leads to a ι, which is smaller or greater than one.
- The mixture of $\iota > 2$ is completely unmixed (like $\iota = 0$).
- The classification includes only perfectly mixed and perfectly unmixed.

The evaluation function v is a function according to the following approach and displayed in Figure 3.37.

$$v = 1 - |1 - \iota| \qquad 3.17$$

This evaluation function v is calculated for each cell and each time step. The evaluation value N is defined as follows and can be titled as the ι weighted accumulated CO mass or oxidation-limit number:

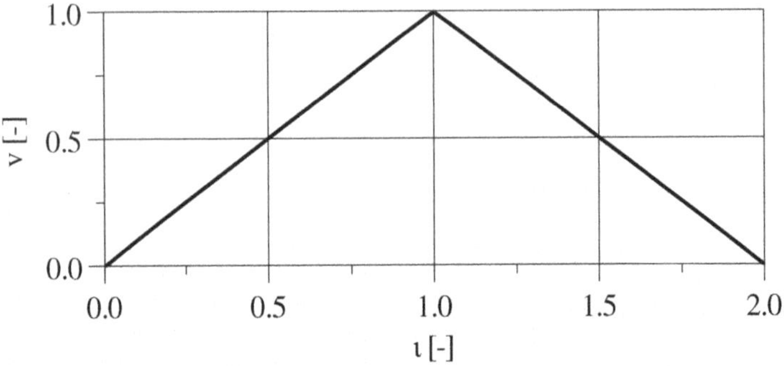

Figure 3.37: Evaluation function v with changing ι.

$$N = \frac{\sum_{i=0}^{m}\sum_{j=1}^{k} m_{CO_{i,j}} v}{\sum_{i=0}^{m}\sum_{j=1}^{k} m_{CO_{i,j}}} \qquad 3.18$$

$k =$ Amount of evaluated cells
$m =$ Amount of evaluated time steps

The closer the local ι is to one, the more heavily weighted is the local mass of CO. If N equals one, the completely unoxidized CO is stoichiometrically mixed. Therefore, the lack of reaction speed is the limiting parameter. If N equals zero, the completely unoxidized CO is unmixed, and the lack of mixing inside the exhaust manifold is the post-oxidations limiting factor since all mixed CO have been oxidized. The diagram in Figure 3.38 shows the desired quantification of the ι distribution for case #1, #2, #3 and #4. While for case #1, the reaction rate and the mixing level seem to have an almost similar influence on the amount of not oxidized CO mass flow, the mixing level dominates the remaining cases.

A more detailed view of the results shows that each influencing factor's importance also depends on the investigated position. The evaluation of N for different positions inside the exhaust manifold is shown in Figure 3.39 (case #2). N decreases starting from a position very close to the exhaust valves up to position 3. This is because the mixed CO oxidizes and the share of the remaining CO is more and more unmixed. Therefore, the mixing level is at position 3 a more dominant limiting factor for the post-oxidation as close to the exhaust

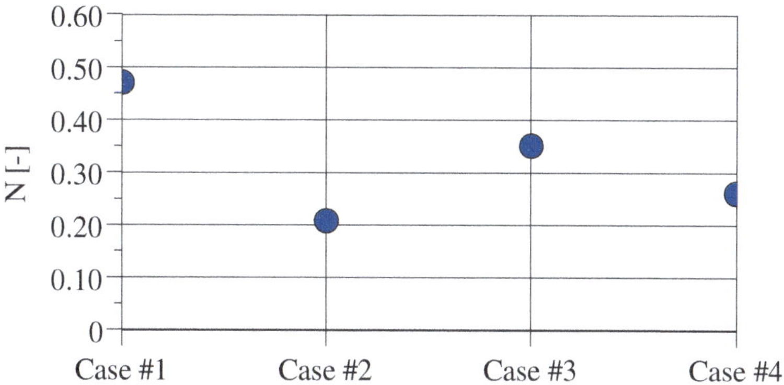

Figure 3.38: Oxidation-limit number N in position 4.

valves (close to position 1). This can be explained by the fact, that close to position 1 all CO is just recently mixed and therefore, the available time span for a reaction start is extremely small. Between position 3 and 4, the mixing increases due to a complex vortex system (as mentioned in chapter 3.2.5). The amount of CO, which is mixed increases and since the available time span for freshly mixed CO between position 3 and 4 is quite small leads to the result that some of the mixed CO is not oxidized until it reaches position 4. This effect is comparable to the conditions close to position 1.

This increase between position 3 and 4 can be observed for all four investigated engine operating points. However, the grade of N at position 3 differs for different engine operating points. This can be seen in Figure 3.40. Case #3 and case #4 show a more dominant mixing dependency than case #1 and case #2. A possible correlation might be as follows. Figure 3.41 shows the dependence of N at position 3 and the engine speed. A higher engine speed limits the available time span to oxidize the mixed exhaust gases. Therefore, the amount of unoxidized but mixed CO mass at position 3 increases with increasing engine speed.

While oxidation-limit number N in between the adaptor shows such a dependence on the engine speed, the N distribution at position 4 is a combination

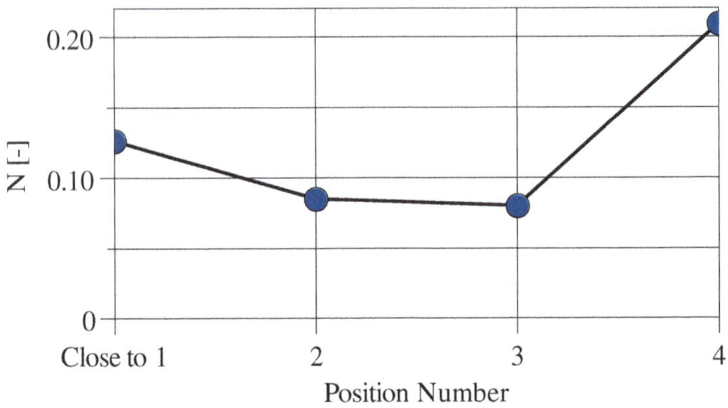

Figure 3.39: Oxidation-limit number *N* in different positions – case #2.

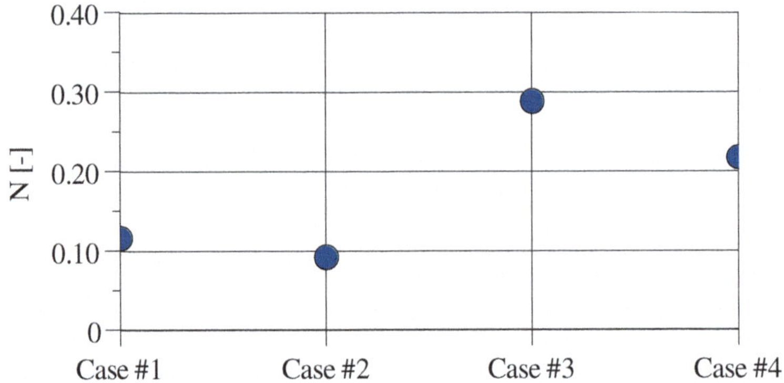

Figure 3.40: Oxidation-limit number *N* in position 3.

of several influencing factors. Beside the engine speed, the dependency of *N* regarding the exhaust gas temperature and the heat flux at the exhaust manifold walls is investigated. This sensitivity analysis is shown in Figure 3.42 and Figure 3.43.

Figure 3.42 illustrates the dependency of *N* regarding the exhaust gas peak temperature. The chosen engine operating point is case #4. The sensitivity of *N* is analyzed with respect to the peak exhaust gas temperature, as described

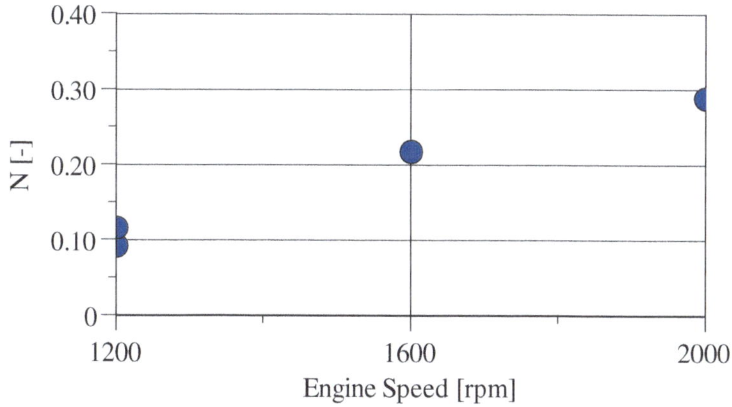

Figure 3.41: Oxidation-limit number N in position 3 concerning engine speed.

in chapter 3.2.2. A decrease of the exhaust gas peak temperature leads to a shift of N towards one and therefore, the reaction rate increases its dominance as a limiting factor for the post-oxidation process. A decrease in temperature leads to lower reaction rates and therefore, the available time span is too short to oxidize for an increasing amount of mixed CO. As described before, the missing self-sustaining effect and the flame extinguishing effects get more dominant with a decreasing temperature, which could explain the nonlinear behavior of the curve displayed in Figure 3.42. A decrease of 100 K affects mainly the reaction rates, while a further decrease also includes the mentioned effects of missing self-sustainment and the flame extinguishing effects.

The change in exhaust manifold wall boundary conditions from adiabatic to a constant wall temperature of 800 K has a similar effect on N as a moderate decrease in temperature due to the heat flux through the exhaust manifold walls (Figure 3.43, case #4 and case #4c).

Similar investigations can be made for the H_2 masses approaching the turbine volute. The evaluation scheme is similar to the CO mass evaluation, but this time the CO masses are replaced with the H_2 masses. The ι calculation remains the same. The oxidation-limit number with respect to H_2 for case #1, #2, #3 and #4 is displayed in Figure 3.44.

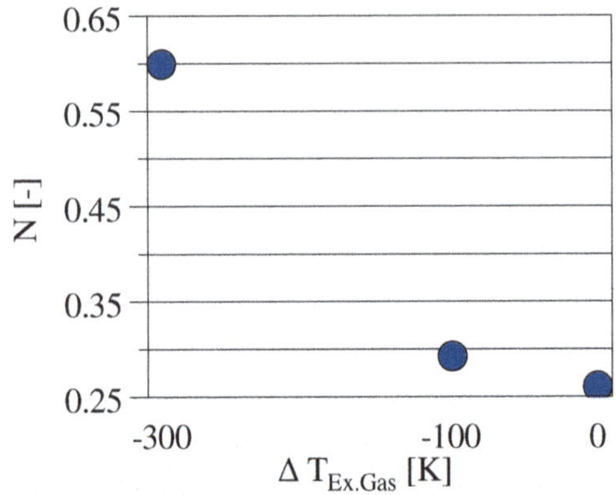

Figure 3.42: Changes of the oxidation-limit number N due to exhaust gas peak temperature sensitivity analysis.

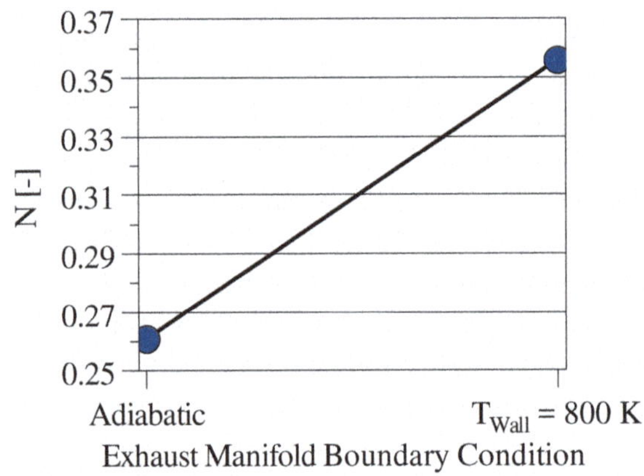

Figure 3.43: Changes of the oxidation-limit number N due to wall boundary condition sensitivity analysis.

The level of N for each case is lower than the corresponding oxidation-limit number with respect to CO. This is due to the higher reaction rate of H_2 at similar boundary conditions in comparison to CO (as explained in chapter 3.2.5). A higher reaction rate leads to a less significant impact of effects like late mixing (between positions 3 and 4).

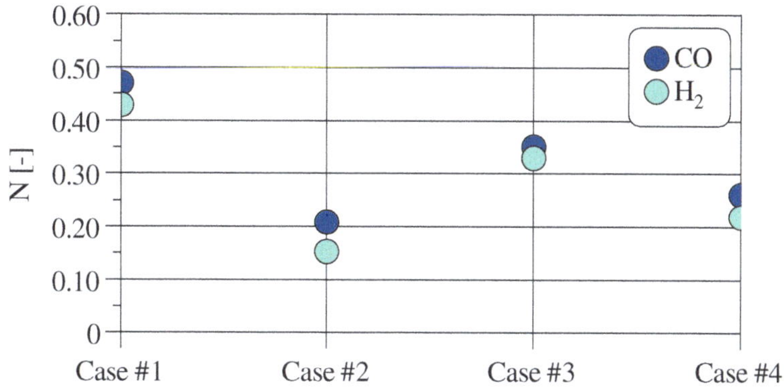

Figure 3.44: Comparison of the oxidation-limit number regarding CO and H_2 at position 4.

The evaluation shows that the mixing level is a more dominant limiting factor for the post-oxidation process for the bulk of investigated engine operating points.

Nevertheless, the reaction rate dependency of the post-oxidation process depends highly on the chosen boundary conditions. A reduced exhaust gas temperature, a higher engine speed, and increased wall heat fluxes are examples of how the reaction rates' influence on the post-oxidation process is increased. The mixing seems to limit the maximum amount of oxidizable exhaust gas mixture. At the same time, the reaction rate defines the amount of oxidizable exhaust gas within limits set by the mixing. A very fast reaction rate would be able to oxidize all mixed exhaust gas but not the unmixed – based on the simplification, that the oxidation effects do not strongly influence the mixing.

Therefore, a 1D post-oxidation model needs to be capable of displaying the mixing inside the exhaust manifold, in particular concerning the limited mixing between position 1 and 3 and the increased mixing level between position 3 and 4. Based on this mixing level, the reaction rates are the second main influencing parameter, which has to be modelled. Its dependence on temperature, heat fluxes and engine speed is a key to developing a post-oxidation model, which can display the heat releases inside the exhaust manifold with changing boundary conditions. Also, a differentiation of the CO oxidation and the H_2 oxidation is also necessary due to the different behavior regarding the influence of mixing level and reaction rates.

Mixing Evaluation Methodology for 3D and 1D Simulation Domains

The mixing investigation made so far do all include radial mixing as well as axial mixing effects. Since this work concentrates on developing a 1D post-oxidation model, an evaluation of the mixing level needs to be done concerning the capabilities of the 1D simulation domain.

The mixing investigation so far is an excellent tool to evaluate the effects of the mixing and the reaction kinetics. However, to rate the mixing inside the exhaust manifold in a way comparable to a 1D mixing model, the radial mixing effects need to be neglected. The comparison requires an evaluation methodology applicable to both 3D-CFD simulation results and 1D simulation results. The following sequence shows an evaluation approach, which fits both simulation domains.

The evaluation approach bases on the ι evaluation approach discussed above. The difference is the evaluated data. While the ι evaluation approach evaluates the conditions of a selected species mass inside each chosen cell and each chosen time step individually, the now discussed approach is limited to the mean value of a 2D surface cut of the 3D-CFD mesh. This approach assigns each position a scalar value, which rates the mixing at this specific position. Since this mixing rating is a surface averaged scalar, the radial mixing effects are neglected, and only the axial mixing of the flow is rated. This simplification allows the comparison to the 1D simulation domain. This chapter aims to develop an appropriate tool to compare the mixing inside a 3D mesh to the results of a 1D engine simulation at specific exhaust manifold positions.

As already discussed, the mixing level of the exhaust gas inside the exhaust manifold changes as it travels through the pipes. This mixing process and the oxidation process inside the exhaust manifold do occur simultaneously. As shown above, some of the mixed combustion products react during this process. The products of these reactions are then not separable from the rest of the fully burnt emissions like H_2O and CO_2. Therefore, the evaluation of a 3D-CFD simulation with implemented reaction kinetics is not suitable to rate the mixing of the emissions and the fresh air inside the exhaust manifold on its own. Besides, the rating of the mixing effects inside a 1D exhaust manifold model cannot include the oxidation effects because of the lack of a post-oxidation model, which is suitable to display a validated CO and H_2 oxidation.

This approach is subject to the assumption that the flow inside the exhaust manifold shows a similar behavior (regarding mixing effects) with and without activated post-oxidation effects. The boundary conditions at the exhaust manifold inlet and outlet are kept the same. This modification of case #2 is named case #2a and also listed in Table 3.4.

The evaluation is based on a parameter called mixing level M, which is defined as follows:

$$M = \frac{m_{(CO+H_2)_{mixed}}}{m_{(CO+H_2)_{total}}} \tag{3.19}$$

$$m_{(CO+H_2)_{total}} = \int_{Cycle_{Start}}^{Cycle_{End}} \dot{m}_{(CO+H_2)_{total}} dt \tag{3.20}$$

$$m_{(CO+H_2)_{mixed}} = \begin{cases} \int_{Cycle_{Start}}^{Cycle_{End}} \dot{m}_{(CO+H_2)_{total}} dt & \forall \iota \geq 1 \\ \int_{Cycle_{Start}}^{Cycle_{End}} \iota \cdot \dot{m}_{(CO+H_2)_{total}} dt & \forall \iota < 1 \end{cases} \tag{3.21}$$

$$m_{(CO+H_2)_{unmixed}} = \begin{cases} 0 & \forall \iota \geq 1 \\ \int_{Cycle_{Start}}^{Cycle_{End}} (1-\iota) \cdot \dot{m}_{(CO+H_2)_{total}} dt & \forall \iota < 1 \end{cases} \tag{3.22}$$

Equation 3.20 describes the accumulated mass of CO and H_2, passing the evaluated position in each time step. As mentioned before, the mass flow is a mean surface value at the evaluated position.

Equation 3.21 describes the accumulated mass of CO and H_2, which is defined as mixed. The definition of mixed or unmixed underlies the following simplification:

- The mass flow passing the evaluation surface is divided into a perfectly mixed (equation 3.21) and a completely unmixed part (equation 3.22).

- If the local surface average ι of the evaluated time step is greater or equals one, enough oxidizers are available for both, CO and H_2, to oxidize completely (equation 3.20).
- If the local surface average ι of the evaluated time step is smaller than one, (1-ι) times the mass flow is declared as unmixed (equation 3.22), while ι times the mass flow is declared as perfectly mixed (equation 3.21).

An additional simplification needs to be done due to the behavior of the exhaust gas inside the exhaust manifold. Close to the exhaust valves, accumulated residual emissions and fresh air remains during the exhaust strokes of the remaining cylinders. The pressure waves force the residual gases to move in a close area.

These alternating mass flows might pass the evaluation surface several times and therefore, would change the resulting mixing level M. In order to avoid an overrating of masses during this period with possibly still changing mixing levels, the surface averaged evaluations in areas, which are influenced by mainly just one exhaust stroke event – the valve area and the adaptor area – are not made for the whole engine cycle but only for the time span with the increased mass flow. Due to the small mass flow rates in time spans beside the exhaust stroke of the directly attached cylinder, the evaluation error would be rather small.

As an example, the evaluation is carried out for case #2a and illustrated in detail. The chosen positions are position 2 and position 4 (displayed in Figure 3.45 and Figure 3.46). Both diagrams display the accumulated mass of CO and H_2 in total according to equation 3.20 and the accumulated mass of the CO and H_2, declared as mixed according to equation 3.21. The third line in each diagram represents the mixing level M during one engine cycle. In Figure 3.45, the evaluation stops at a crank angle of -320 °CA for the reason explained above. In case #2a, the mixing level M is approximately 12 % at position 2 and approximately 44 % in position 4. This procedure can be applied to every chosen position in the exhaust manifold mesh and any chosen engine operating point.

Figure 3.47 shows the resulting mixing level M for both available engine operating points without implemented reaction mechanism, each including

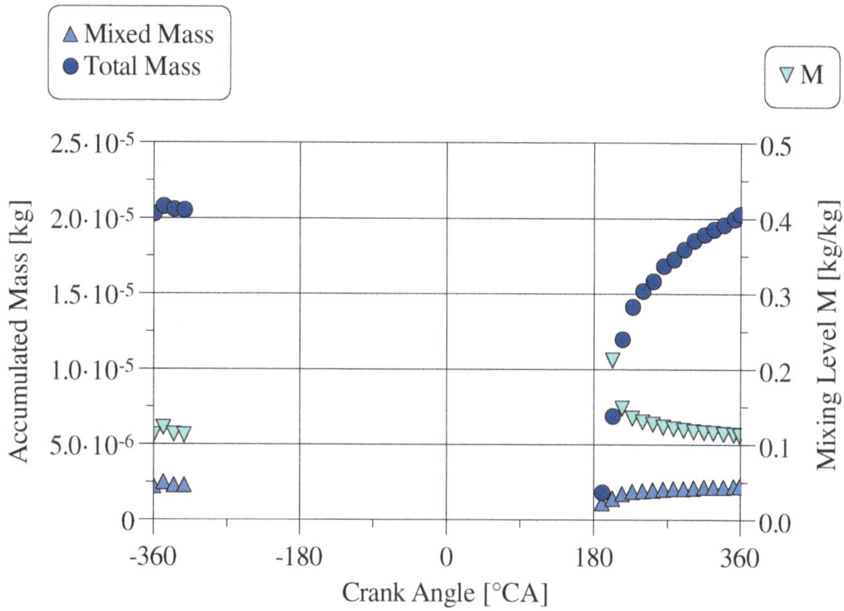

Figure 3.45: Accumulated combined mass of CO and H_2 mixed and unmixed with the mixing level M in position 2 – case #2a.

three evaluation positions. The diagram shows that the basic behavior of the two engine operating point concerning mixing is quite similar. The mixing is limited in the valve and adaptor area and increases between positions 3 and 4. This finding resembles the findings made in chapter 3.2.6. The difference in this investigation in comparison to chapter 3.2.6 is the missing reaction mechanism. Therefore, the mixed CO and H_2 inside the exhaust manifold can be tracked until the turbine without considering already burnt CO and H_2.

The mixing distribution allows the following assumptions about its origins: The rather limited mixing in the valve area and the adaptor area is due to the geometry of this area. Between position 2 and position 3, the mixing increases in a very restricted way. The geometry at this position is a single pipe (for each cylinder) with a constant cross-section. This inhibits the formation of large eddies and the related mixing of emissions and fresh air. The initial mixing

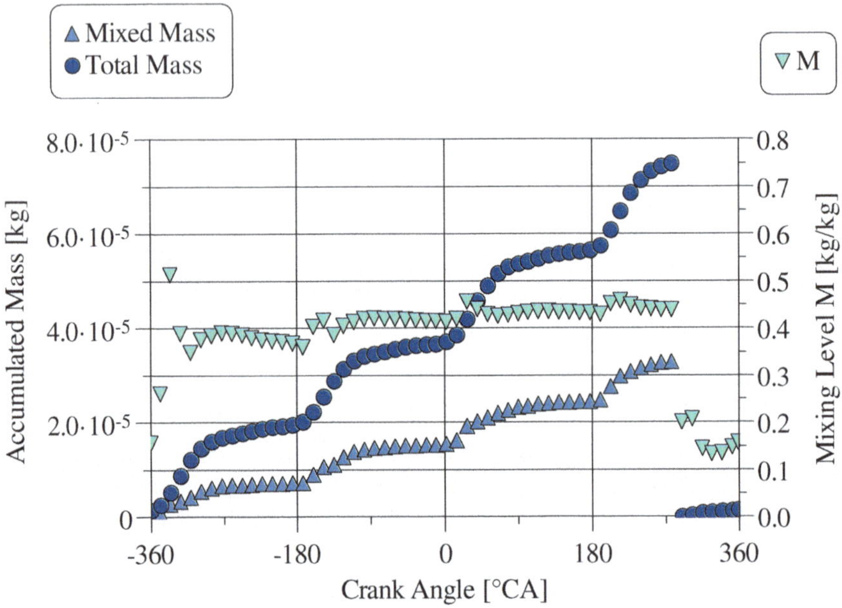

Figure 3.46: Accumulated combined mass of CO and H_2 mixed and unmixed with the mixing level M in position 4 – case #2a.

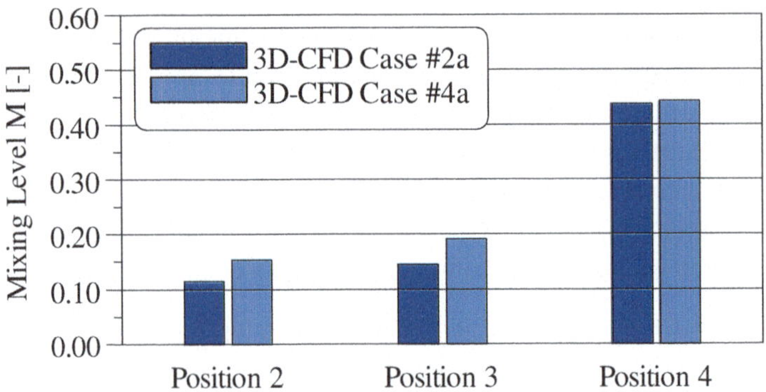

Figure 3.47: Mixing level in different positions (3D-CFD simulation without reaction mechanism).

level at position 2 results from the valve area geometry and the impact of the fast cylinder emissions on the accumulated residual fresh air directly behind the exhaust valve. The valve area geometry, as shown for example in Figure 3.19, contains a pipe junction. This junction might have a mixing promoting effect on the flow. The second and most certainly, the more important factor is the emission flow's impact with a rather high speed and the accumulated stationary residual fresh air.

Comparing the two evaluated cases in Figure 3.47 displays the difference in the mixing level at position 2. This difference might have two reasons. The first reason is the different emission rate at the exhaust valve and the related emission speed. This deviation is due to the different engine speed of the two cases. A higher engine speed leads to a higher mixing in the valve area. A second factor is the amount of incomplete burnt emissions, which is entering the valve area. Although the scavenging air mass is proportional to the amount of incomplete burnt emissions (this is a consequence of the stoichiometric engine operation), the pipe cross-section remains constant. Thus, the interaction zone size of the exhaust gas and the fresh air is not directly proportional to the amount of incomplete burnt emissions.

The increase of the mixing level between position 3 and position 4 matches the findings displayed in Figure 3.23 and Figure 3.39. The mixing level is higher in the mixing and the turbine area, and so is the volume-specific heat release. The assumption of increasing mixing between positions 3 and 4 can now be evaluated by using the mixing level M.

3.3 Summary and Discussion

In this chapter, the post-oxidation effect is investigated in great detail. In section 3.1, the micro-level investigation of the post-oxidation effect is discussed. The oxidation process, especially the chemical ignition delay of CO and H_2, highly depends on the radical formation as an intermediate product. Since the oxidation process inside the exhaust manifold is, due to the limited residence time of the exhaust gas, time restricted, the chemical ignition delay plays a major role in these reaction kinetic evaluations. The chemical ignition delay increases with changing boundary conditions that inhibit the formation of OH-radicals. For example boundary conditions that favor the HO_2 formation. On the other hand, a reduction of the chemical ignition delay can be achieved with measures that favor the OH-radical formation, as for example the addition of NO at lean combustion conditions.

The evaluation of the 3D-CFD simulation results shows a complex interaction of boundary conditions on the chemical heat release inside the exhaust manifold due to the post-oxidation process. The volume specific heat release of each defined exhaust manifold area (Figure 3.23) illustrates the importance of the reaction speed on the oxidation process. The chemical ignition delay and the self-reinforcing effect are the key elements to describe this behavior.

Beside the reaction rate, the mixing inside the exhaust manifold is of great importance. Since only the exhaust gas that is mixed up with additional fresh air is able to oxidize, the mixing limits the maximum amount of releasable heat inside the exhaust manifold. The mixing is a highly three-dimensional effect that is mainly affected by the exhaust manifolds geometry. The pronounced mixing in the mixing area is due to the increasing pipe cross-section and the subsequent flow separation. Areas with rather constant pipe cross-section inhibit a notable vortex formation, leading to a limited mixing (especially the adaptor area). The heat release in the valve and adaptor areas are mainly the result of the mixing that occurs due to the interaction of fast pushed out exhaust gas and the accumulated scavenging air right downstream of the exhaust valves.

The interaction of reaction kinetics and mixing effects and their effect on the oxidation of CO and H_2 during an entire engine cycle needs to be evaluated. Due to the absence of an useful evaluation tool that can be applied to the 3D-CFD simulation results, the oxidation-limit number N is developed. The oxidation-limit number is based on ι. This ratio rates the local availability of oxidizers related to the local CO and H_2. In each cell at a specific position in the evaluated mesh, the approaching CO and H_2 that is not yet oxidized, is evaluated regarding ι for each time step of an entire engine cycle. The CO or H_2 masses are then sorted according to their individual ι, using a histogram.

The resulting histogram is evaluated using an evaluation function that rates unmixed masses (or masses with way too many oxidizers) with a factor of zero and stoichiometrically mixed masses with a factor of one. For mixtures between these extreme values, linear interpolation is used. The resulting rated masses are accumulated and normalized.

The result is the oxidation-limit number N that indicates the condition of not oxidized emissions at the evaluation position. The closer N gets to one, the better mixed the unoxidized exhaust gas. From this it can be concluded, that the slow oxidation rate of the exhaust gas is the reason for the limited oxidation. An oxidation-limit number close to zero indicates that the bulk of unoxidized exhaust gas is characterized by the absence of oxidizers or the availability of way too much oxidizers. From this it can be concluded that the well mixed exhaust gas is already burnt on its way to the evaluation position and the limited mixing limits the oxidation process inside the exhaust manifold.

Using this new developed 3D-CFD simulation evaluation tool, the condition of the exhaust gas can be evaluated at each position inside the exhaust manifold.

The investigation shows that both, mixing and reaction kinetics, are relevant for the post-oxidation process inside the exhaust manifold. Their individual importance is affected by the engine operating point. Therefore, both effects need to be modelled to obtain a 1D post-oxidation model.

To rate the mixing inside the exhaust manifold in a way that can be compared with a 1D simulation approach of the engine, a different evaluation tool is necessary since N relies on 3D-CFD simulation results that display axial as well as radial mixing effects. In addition, the evaluated simulations do not include a reaction mechanism, since the mixing behavior must not be overlaid by oxidation effects.

An evaluation methodology applicable to both simulation domains is the mixing level M. This approach neglects the radial mixing effects of the 3D-CFD simulation results by surface averaging the local species concentration at each evaluated position. The result is a scalar value of each species' concentration and mass at each evaluated position for both simulation domains, 3D-CFD and 1D simulation. The surface averaged species concentration is evaluated for each time step of the simulation individually.

Based on the local ι the CO and H_2 masses can be separated into stoichiometrically mixed and unmixed. The accumulated mixed mass of CO and H_2 during one entire engine cycle is normalized by the total mass of the engine cycle. The resulting ratio shows the share of mixed CO and H_2 at this specific position for the entire engine cycle.

It is clear that this evaluation methodology includes a wide range of simplifications but since the 1D model approach is also characterized by many simplifications regarding flow field modelling inside the exhaust manifold, this approach is considered sufficient. The mixing level M allows a comparison of the species behavior inside an exhaust manifold for 3D-CFD and 1D simulation domains.

4 Post-Oxidation Model Development

The findings so far show the complexity of the post-oxidation process. Three-dimensional effects and a complex reaction kinetic characterize it. In the following, a 1D modelling approach is investigated. The 1D engine simulation is made by usage of the simulation software GT-POWER (Gamma Technologies [2]).

The underlying 1D engine model bases on the test-bench engine mentioned in chapter 3. The model is calibrated with the test-bench engine experimental data using a Three-Pressure-Analysis and a subsequent combustion modelling using Wiebe functions. The engine model includes the 10 cm adaptors downstream of each of the four cylinders. These adaptors are necessary to include sensors at the test-bench. In order to ensure comparability, these adaptors are also modelled within the used 1D engine model. More details on the test-bench engine and the 1D model calibration can be found in [1].

This chapter addresses the development of a 1D post-oxidation model. Based on the findings of the 3D-CFD evaluation, the finished model should be able to represent not only the heat release due to post-oxidation but also the formation of CO_2 and H_2O with a simultaneous decrease in the CO, H_2 and O_2 concentration. The heat release should be reasonably approximated both in its absolute expression and in its local distribution.

As already pointed out in chapter 3.2.6, both mixing and reaction rates have a major influence on post-oxidation. Whether the mixing or the reaction rate has a greater limiting influence on post-oxidation depends on the respective boundary conditions in the engine exhaust manifold. Therefore, these influences must not only be reflected in the model but must also respond in a meaningful way to the changes in boundary conditions. The effects of mixing and oxidation are modelled separately.

The modelling of mixing effects is based on the operating points presented in chapter 3.2.6 without an implemented reaction mechanism. This allows a comparison of the mixing conditions in the entire exhaust manifold. This

J. Przewlocki, *Simulative Investigation of Post-Oxidation in the Exhaust Manifold of SI Engines*, Wissenschaftliche Reihe Fahrzeugtechnik Universität Stuttgart,
https://doi.org/10.1007/978-3-658-36378-9_4

comparison at different positions in the exhaust manifold is essential because the exhaust manifold can be divided into two sections. Due to their geometry, the valve and adaptor areas tend to be low in mixing, while the mixing and turbine areas are high in mixing. Since mixing determines the maximum oxidizable fraction of incomplete burnt exhaust gas, these different areas within the exhaust manifold are immensely important for a functioning post-oxidation model.

Oxidation, in turn, determines how much of the mixed exhaust gas actually oxidizes before reaching the turbine volute. Oxidation is modelled using a reaction rate that responds to temperature changes based on the modified Arrhenius equations. Here, the direct implementation of the previously used reaction mechanism is contrasted with a simplified approach with only two reaction equations.

The two sub-models, the mixing and the oxidation model, are then implemented in the 1D engine model mentioned above.

4.1 Comparison of 3D-CFD and 1D Boundary Conditions

The post-oxidation process inside the exhaust manifold heavily depends on the conditions of the incomplete burnt exhaust gas. A higher absolute amount of incomplete burnt exhaust gas leads, as described in chapter 3.2.6, to a lower percentage mixing, since the amount of scavenging air also increases but the pipe cross-section remains the same. Beside the absolute amount of species, the corresponding temperature is a relevant influencing parameter. This chapter compares the species mass entering the exhaust manifold in each cycle and the corresponding temperatures of the emissions during the exhaust stroke. The emissions of the in-cylinder combustion of the 1D engine model can differ from the 3D-CFD emission boundary conditions due to the difference in combustion calculation.

According to chapter 2.3.2, the 1D engine models species composition of the exhaust gas mainly depends on the pressure and temperature. A model tuning – to maintain similar emissions in comparison with the 3D-CFD boundary

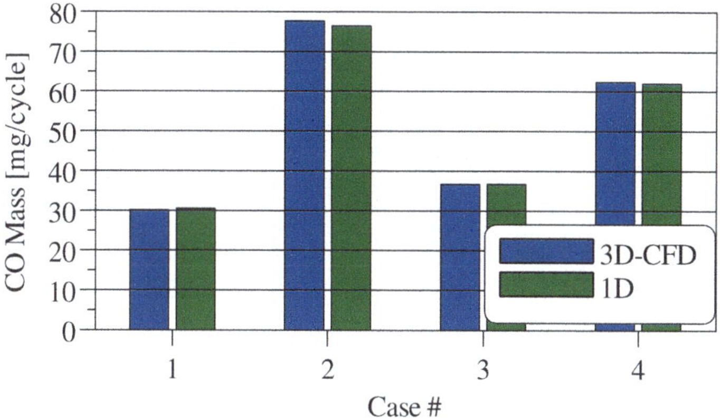

Figure 4.1: Comparison of the 3D-CFD and 1D simulation *CO* emissions of all four cylinders.

conditions at the exhaust manifold inlet – can be achieved by a change of the center of combustion. Thus, it is possible, that either the emissions of the 1D simulation do not match the emissions out of the 3D-CFD simulation or the corresponding temperatures do not match. In this case, the species composition is chosen to be the more important parameter and the center of combustion is edited in a way so that similar emissions are achieved. This weighting is made since the first step is to investigate the mixing behavior of the emissions inside the exhaust manifold. The amount of incoming emissions is a major influence factor. The temperature is more important in terms of the oxidation process, which is the second step.

The absolute masses of each, CO and H_2 emissions during one engine cycle are displayed in Figure 4.1 and Figure 4.2 (total mass of all four cylinders). The masses are about to be similar to the 3D-CFD simulation results. The center of combustion of each cylinder is slightly tuned in order to match the emission as displayed in the diagrams. The relative difference of the exhaust gas peak temperature is displayed in Figure 4.3 (in each case, cylinder one is chosen to be compared). The maximum difference in exhaust gas peak temperature is -7 %. The diagram shows that the tuning of the center of combustion does not lead to an intolerable discrepancy in temperature. Both, the exhaust gas

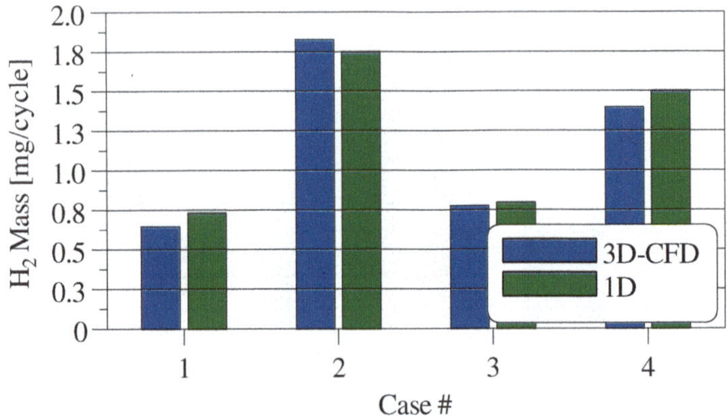

Figure 4.2: Comparison of the 3D-CFD and 1D simulation H_2 emissions of all four cylinders.

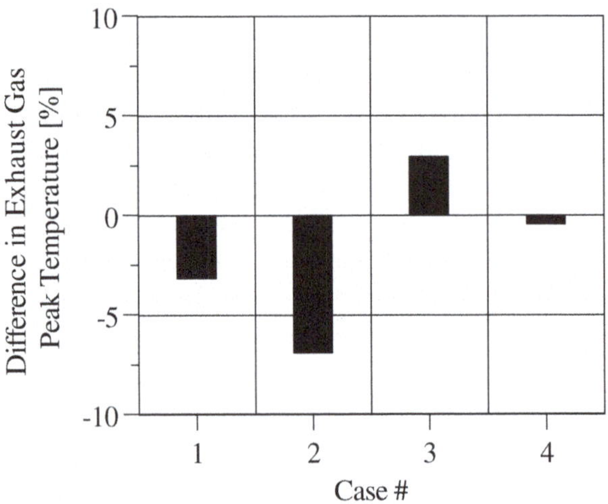

Figure 4.3: Relative difference of the exhaust gas peak temperature of 3D-CFD and 1D simulation.

temperature and the emission mass out of the 1D model match the 3D-CFD results quite well.

4.2 Modelling of Mixing Effects

The 1D post-oxidation model relies on the 1D engine flow solver GT-POWER [2]. Inside a 1D engine model, the flow model solves the Navier-Stokes equations in one dimension [2]. Solving a 1D Navier-Stokes equation implies an averaging of the flow quantities of the flow field perpendicular to the flow direction. As a result, it is impossible to dissolve the flow field in another dimension than the flow direction dimension. The resolution of the flow field in flow direction relies on the discretization length. The smaller the discretization length, the finer is the resolution of the flow field. A finer resolution leads to a smaller subvolume, in which the flow is averaged into a single scalar. Beside a finer resolution, the smaller discretization length leads to an increased calculation expense because of the necessity to solve the Navier-Stokes equations in every subvolume created by the discretization length. In addition, if an explicit flow solver is used, the corresponding time step, which is necessary to maintain a stable model behavior, decreases. This increases the amount of time steps in which the Navier-Stokes equation must be solved in each case. The 1D engine model is based on an explicit flow solver due to its capability to resolve small time steps within a reasonable time. These small time steps are necessary to display effects like pressure pulsations and highly resolved species concentration gradients within the flow.

The resolution of just those species concentration gradients is the key to the 1D flow mixing modelling. A large discretization of the modelled flow components like a pipe results in a limited amount of subvolumes, which resolve this very pipe. In a coarsely resolved pipe, single separated species clouds are always averaged within the same subvolume. If the separated species clouds are inside a fine resolved pipe, the share of the clouds, which are both inside the same subvolume, decreases compared to the coarsely resolved pipe. Then, larger amounts of the clouds are not averaged except for a small share of each cloud.

As already mentioned, a finer resolution of the flow components leads to an increased computation time of the engine model. Mostly, this is avoided by choice of a rather large discretization length inside the exhaust manifold. This procedure lowers the computation time while pressure pulsation effects can still be resolved.

In the following, the 1D engine model, as calibrated in [1] and briefly discussed in chapter 4, is used, and the mixing inside the 1D engine model exhaust manifold is investigated. The evaluation is made according to the methodology developed in chapter 3.2.6. The evaluation positions are chosen to be similar to the positions defined in Table 3.7. In contrast to the 3D-CFD evaluation, the evaluated parameters do not have to be surface averaged due to the characteristics of the 1D simulation (as explained above). The engine operating point of choice is case #2a. This case is chosen because of the deactivated reaction mechanism. This allows a comparison of the 3D and 1D simulation results. The second available engine operating point without an activated reaction mechanism, case #4a, will be used for mixing model validation.

The baseline model has a rather coarse discretization inside the exhaust manifold in order to keep the computation time low. This leads to a suboptimal resolution of species concentration and in addition to that, the spatial distribution differs from the 3D-CFD mesh. Figure 4.4 shows the volume distribution of the baseline model in comparison to the 3D-CFD mesh. While the valve area is smaller, all remaining areas have each a higher volume in comparison to the 3D-CFD mesh. Consequently, the overall volume of the 1D engine is greater than the volume of the 3D-CFD mesh. It is important to stress that a different exhaust manifold volume is not state of the art, but it is in this example, the volume needs to be adjusted.

A second important difference of the baseline 1D exhaust manifold in comparison to the 3D-CFD mesh is the modelling of the mixing area. The 3D-CFD mesh shows differences for the different cylinders. Cylinders number one and four are located at the two outer positions. Therefore, the gas flowing out of them must cover a longer distance on its way to the turbine volute. This increases the distance of the fluid to an area with increasing pipe cross-section or increasing volume in general. This geometric feature of the 3D-CFD mesh is necessary to develop the vortex system responsible for the mixing in the mixing zone. The baseline 1D exhaust manifold does not include such a difference in distance of the cylinders.

Besides, the pipes that represent this distance between the adaptor and the pipe junction are very short. In compensation to that, the junction of the four pipes and the pipe leading to the turbine is designed as a single volume with a volume of 0.7 l. Figure 4.4 shows that this single volume makes up almost the entire mixing zone volume. The design as a single volume leads to a similar averaging as a single subvolume of a pipe. Therefore, the single volume pipe junction results in a huge mixing effect in species concentration averaging.

Figure 4.6 shows the mixing level of the baseline model for case #2a. Besides the just discussed effect of the single volume pipe junction (mixing level between position 3 and 4 doubles), the effect of the coarse discretization of the pipes is observable. The mixing level inside the adaptor increases from approximately 15 % up to 25 %. The diagram also includes the results of the 3D-CFD evaluation. A comparison with the baseline model evaluation shows a too high mixing level of the 1D baseline model at position 4 as well as during the total travelling distance of the fluid (the evaluation is made for the exhaust gas out of cylinder 1).

Subsequent, the geometry of the 1D engine model exhaust manifold is modified. The modification is split up in two individual steps in order to be able to consider the effects of the measures separately.

The first modification includes the rearrangement of the exhaust manifolds volume. This includes the overall volume as well as the volume distribution according to the 3D-CFD exhaust manifold mesh. The rearrangement also includes the individual pipe length of each cylinders flow path. The already mentioned single pipe volume representing the pipe junction in the mixing area is downsized from 0.7 l to 0.07 l. The resulting mixing level is displayed in Figure 4.6. The rearrangement of the volume distribution increases the mixing level at position 2 due to the increased volume of the valve area. The decrease of the pipe junction in the mixing area limits the mixing by 10 % compared to the baseline setup.

The mixing level is still too high in comparison to the 3D-CFD results. A further decrease is made with the second modification.

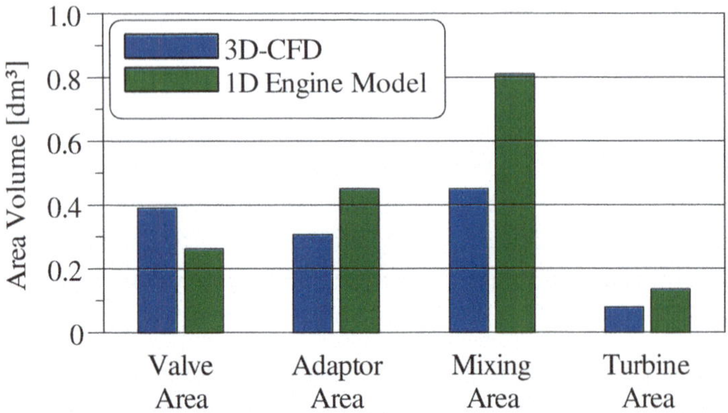

Figure 4.4: Average area volume of the 3D mesh and the not modified 1D engine model.

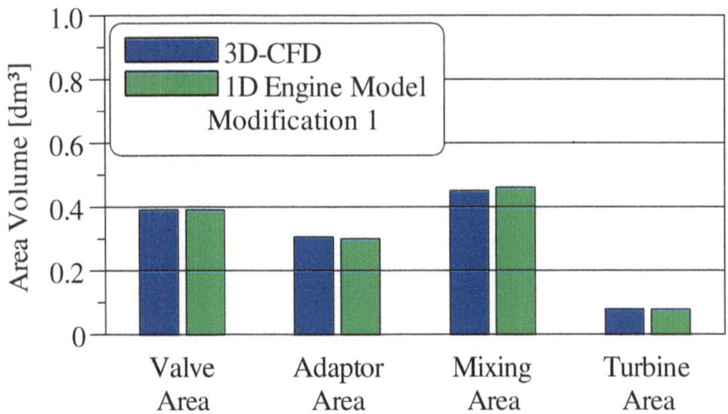

Figure 4.5: Average area volume of the 3D mesh and the modified 1D engine model.

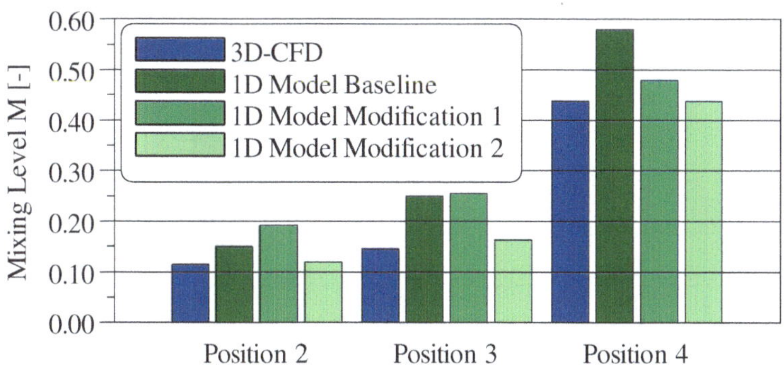

Figure 4.6: Mixing level of different positions of the 3D-CFD simulation result and 1D simulation results inclusive modifications of the exhaust manifold modelling.

This modification includes the decrease of the discretization length of the pipes inside the exhaust manifold. The discretization size is made up to six times smaller. This procedure leads to less averaging inside the pipes in the exhaust manifold from which in turn follows a reduced mixing inside these pipes. The mixing does now fit the mixing level of the 3D-CFD results very well.

With these modifications, the 1D engine model is capable of reproducing the mixing level inside an exhaust manifold, including scavenging conditions. The second available 3D-CFD simulation without implemented reaction mechanism is now used as a validation case (case #4a, as listed in Table 3.4). Figure 4.7 shows the mixing level of case #4a for both 3D-CFD simulation and 1D engine model simulation. Again, the mixing level shows a limited mixing up to position 3 and an increased mixing between positions 3 and 4. The 1D engine model simulation results display a similar mixing level distribution inside the exhaust manifold. The mixing level at position 2 and 3 is also limited – on a lower standard than with modification 1 – and increases between position 3 and 4 up to 45 %. Again, the 1D engine model simulation shows a quite similar mixing level at the defined positions with a difference at position 2 and 3 of up to 5 %. Based on these results, it is assumed in the following, that the mixing level inside the exhaust manifold of the 1D engine model is approximately comparable to the mixing level of the 3D-CFD results. The most

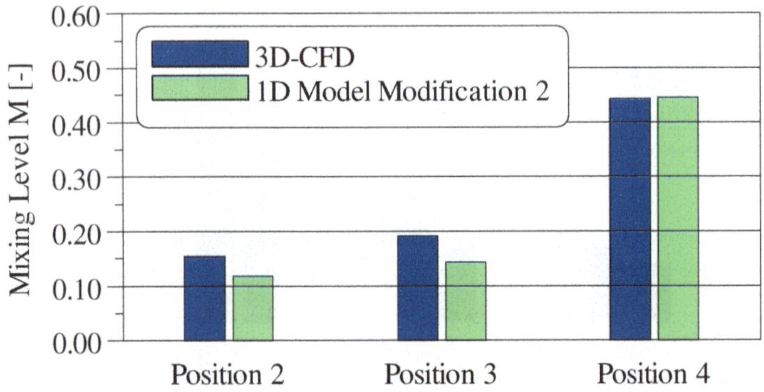

Figure 4.7: Mixing level of different positions of the 3D-CFD simulation result and 1D simulation results (validation case #4a).

relevant tendencies, namely the low mixing level up to position 3 and the highly increased mixing level at position 4, can be reproduced. It is, of course, clear, that the modelling of three-dimensional mixing effects like a vortex due to flow separation is not possible with the usage of a 1D engines flow model. The target of the 1D engine mixing model is to be able to display the characteristics of the exhaust manifolds mixing in general. On the one hand, this means a low mixing level in areas with relatively constant pipe cross-sections. On the other hand, the mixing level increases after areas characterized by increasing pipe cross-sections and increased three-dimensional flow behavior.

The distribution of the mixing level is reproducible, but the origin is different. The low mixing in the valve area and the adaptor area is realized by using very small pipe junctions and small discretization lengths. The initial mixing of the exhaust gas approaching the accumulated scavenging air is modelled with the single volume pipe junction connecting the two exhaust valves flows. The increased mixing in the mixing area and the turbine area is achieved with the usage of a pipe junction that has a bigger volume as the subvolumes of each pipe.

In addition to the mixing level inside the exhaust manifold, the accumulation of residual scavenging air close to the exhaust valves is a characteristic flow behavior according to the 3D-CFD results. As already mentioned in chapter 3.2.5,

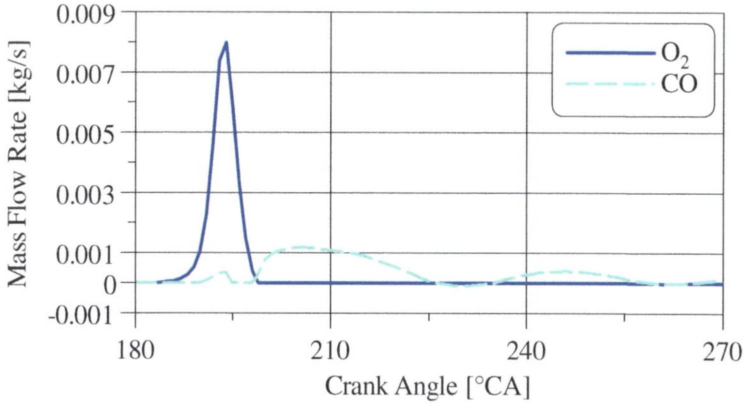

Figure 4.8: 1D simulation results: Mass flow rate inside the exhaust manifold in position 3, case #2a.

instead of flowing directly into the turbines, the scavenging air flowing through the exhaust valve is pushed back to the exhaust valves due to the following cylinder's exhaust stroke. The exhaust gas of the following cylinder is then pushed into the residual scavenging air of the cycle before. Due to this effect, the mixing of the incomplete burnt exhaust gas and the residual scavenging air is assured. Additionally, the accumulated scavenging air can be heated up, if the boundary condition of the exhaust manifold walls is not defined as adiabatic (as discussed in chapter 3.2.4).

Figure 4.8 displays the GT-POWER simulation results of case #2a with the engine model, including adjusted mixing (as described above). The evaluation position is position 3. The diagram displays the mass flow rate of O_2 (blue line) and CO (red line) for cylinder one and one entire engine cycle. The principle of operation of scavenging leads to the fact that the scavenging air follows the exhaust gas. This principle is defined by the valve timings of the engine. In Figure 4.8, the scavenging air (represented by O_2) approaches position 3 before the exhaust gas (represented by CO). This change in order is due to the accumulation of scavenging air (and therefore O_2) close to the exhaust valve. Therefore, the 1D engine model is capable of displaying this phenomenon inside the exhaust manifold.

4.3 Modelling of the Oxidation Effects

An appropriate and – considering the simplifications – realistic described mixing of incomplete burnt exhaust gas with fresh scavenging air inside the exhaust manifold is the base of a working post-oxidation model. The mixing defines the maximum amount of releasable heat inside the exhaust manifold for each defined position. As shown in chapter 3.2.6 besides the mixing, the reaction rate of the oxidation reactions is the second important factor of the post-oxidation process. The reaction rates describe the heat release quantity that is actually released within limits set by the mixing level. The heat release quantity depends mainly on the mass flow temperature. The temperature of the fluid again is influenced by the initial exhaust gas temperature, the heat fluxes at the exhaust manifold walls and by the heat release of the post-oxidation itself.

Therefore, the oxidation model needs to display, beside the conversion of species (from educts to products), the released heat due to the oxidation process, the change in temperature and the changing reaction rate itself due to this change in temperature.

The oxidation model must be implemented into the modified 1D engine model, which is the outcome of the mixing modelling in chapter 4.2. Consequently, the oxidation model development must be done in a way, which is implementable into this 1D engine model. Since GT-POWER [2] is used as 1D engine simulation software, the first step is to evaluate the possibilities of this software.

GT-POWER offers the possibility to affect the 1D flow inside the exhaust manifold by means of a reaction mechanism. The reaction mechanism can be coupled with the flow components of the 1D models exhaust manifold using a predefined chemical reaction template. Inside a flow component, connected to the chemical reaction template, each subvolume includes, besides the in chapter 4.2 implied differential equations, an equation system dedicated to the used reaction mechanism. The size of the additional equation system depends on the amount of species and reaction equations of the used reaction mechanism. The number of additional equations to solve directly affects the necessary computation time.

The simulation software can then display the change in species concentration and heat release inside each connected flow component based on the reaction rates defined by the chemical reaction template.

The parameters that define participating species, the reaction equations and the reaction rates need to be defined in this chemical reaction template. The definition of the reaction rate depends on the local species concentration and on the local temperature. The temperature dependency of the reaction rate is defined by the usage of the modified Arrhenius equation.

As the mixing effects of the investigated exhaust manifold are displayed in a simplified but satisfying way, the most straight forward way of modelling the reaction kinetics inside the exhaust manifold is the implementation of the same reaction mechanism as used for the 3D-CFD simulations (reaction mechanism 2 [12], [27], listed in Table 3.2).

4.3.1 Direct Reaction Mechanism Implementation Approach

In the following, the reaction mechanism, according to [12] and [27] is implemented into the 1D engine model, and the results are evaluated and discussed. As mentioned above, the idea is obvious that the same reaction mechanism combined with a similar mixing level inside the exhaust manifold leads to similar results regarding post-oxidation. With similar amounts of emissions and similar exhaust gas temperatures, the reaction kinetic behavior of the 3D-CFD simulation and the 1D simulation should also be similar concerning post-oxidation rate, chemical heat release distribution and sensitivity analysis. In the following, this procedure is named "1D Reaction Mechanism".

Reaction Mechanism Implementation

The implementation of the chemical reaction template is made according to the schematic illustration in Figure 4.9. In this illustration, the exhaust manifold is roughly sketched by means of the different defined areas. The reaction equations of the used reaction mechanism affect the flow components inside the chosen areas. These chosen areas equal the defined evaluation areas of the 3D-CFD simulation evaluation as defined in chapter 3.2.4.

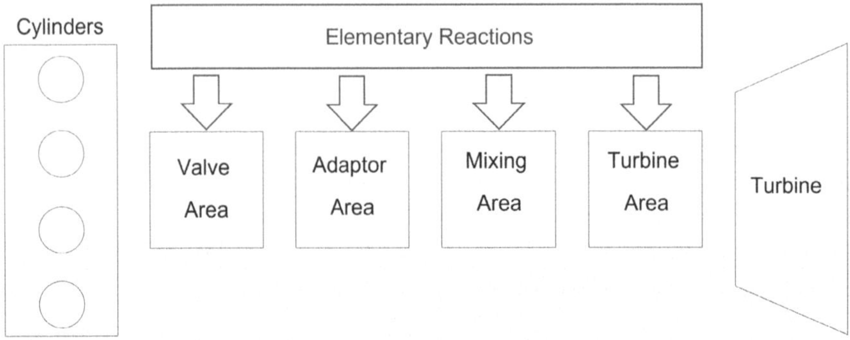

Figure 4.9: Schematic illustration of the 1D engine model with applied chemical reaction template.

The heat release inside the exhaust manifold leads to a different turbine operating point and therefore to a slightly different engine operating point. Hence, this chapter is dedicated to comparing the 1D engine simulation and the 3D-CFD simulation results, the boundary conditions need to be as similar as possible.

Therefore, the 1D engine model is modified as displayed in Figure 4.9. The exhaust manifold is simulated without the engine intake manifold and the cylinders. The turbine and the following parts are also left out. Instead, the boundary conditions are imposed.

This prevents the boundary conditions to change due to the post-oxidation effects inside the exhaust manifold and maintains the comparability of the 1D and 3D-CFD simulations.

Results and Discussion

The results of the 1D engine simulation in comparison to the 3D-CFD results is shown in Figure 4.10. The diagram shows both the post-oxidation rate (as defined in chapter 3.2.4) and the heat release during one entire engine cycle. The results of the 1D reaction mechanism approach show a good agreement compared to the 3D-CFD results. The post-oxidation rate differs by only 15 % (related to the maximum releasable heat).

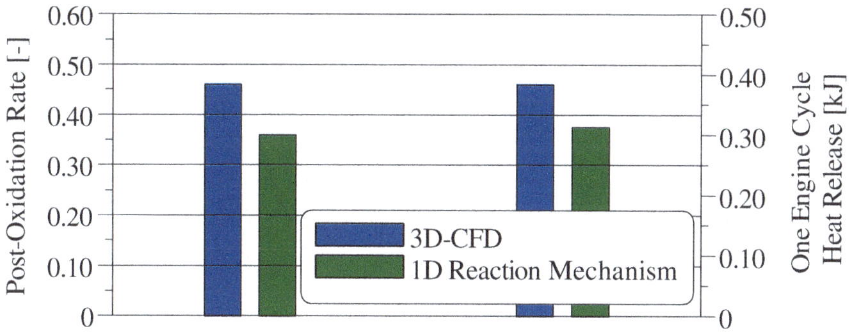

Figure 4.10: Difference in post-oxidation rate and chemical heat release of the 3D-CFD simulations and the 1D model approach including the same reaction mechanism.

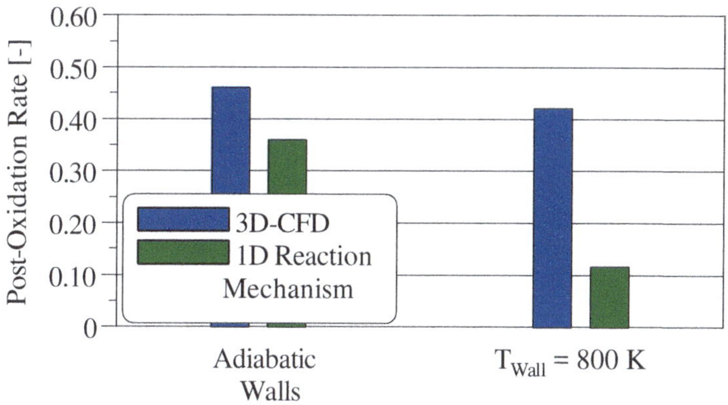

Figure 4.11: Difference in post-oxidation rate of the 3D-CFD simulations and the 1D model approach including the same reaction mechanism concerning changing wall boundary conditions.

In Figure 4.11, the results of a second simulation with changed boundary conditions are shown. According to case #4c in Table 3.5, the exhaust manifold wall temperature is chosen to be 800 K, and the walls are no longer adiabatic. The diagram shows the results of the 3D-CFD simulations again as already discussed in chapter 3.2.4. The 1D simulation results with reaction mechanism show a decrease in the post-oxidation rate down to approximately 8 %, while the 3D-CFD results show a decrease from approximately 46 % to approximately 41 %.

Although the 1D simulation is capable of displaying the heat release of the post-oxidation in an exhaust manifold with adiabatic walls, the results differ quite far if the wall boundary condition is changed to a fixed value. A possible reason for this behavior is the flow modelling of the 1D domain itself. The fixed wall temperatures define a heat flux through the walls based on the difference in temperature of the walls and the fluid. An increased difference in temperature leads to increased heat flux. The direction of the heat flux depends on whether the wall temperature is higher than the temperature of the fluid (heat transfer into the fluid) or whether the wall temperature is lower than the temperature of the fluid (heat transfer from the fluid to the walls and therefore to the environment). The 1D flow model is discretized in the flow direction dimension. This means the discretization is only an axial discretization. The 3D-CFD domain does additionally allow a discretization perpendicular to the flow direction and therefore a radial discretization.

The post-oxidation process is a self-reinforced process. In areas with increased temperature, the heat release rate increases, leading to an even more increased temperature. If the temperature distribution of the pipe cross-section is considered, the area with the highest temperature is not necessarily close to the walls and therefore, the further increase in temperature is not limited. The 1D flow model is not able to display this temperature distribution of the pipe cross-section. The heat flux at the exhaust manifold walls does affect the entire pipe cross-sections mean temperature. This leads to the effect that the wall heat flux always affects the post-oxidation and therefore, always limits the process. In the 3D-CFD simulation case, the post-oxidation process is limited if the flow field inside the pipe causes the high temperature areas to be pushed towards the wall.

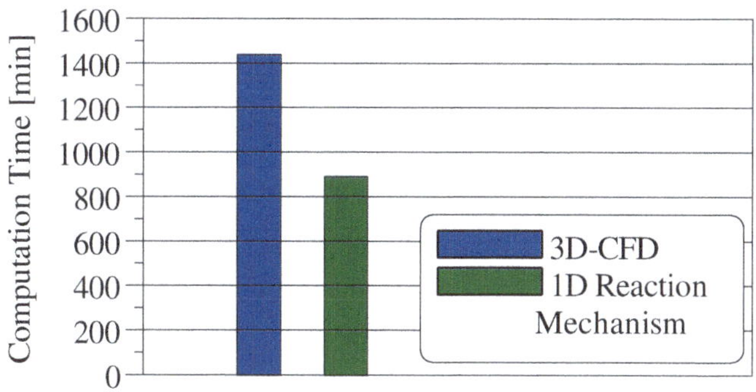

H

Figure 4.12: Difference in computation time of the 3D-CFD simulation of a single case and the 1D engine simulation of a single case with the same reaction mechanism being implemented.

This different behavior due to a missing radial distribution of the temperature might lead to a different heat flux at the exhaust manifold walls. Even if the heat flux at the exhaust manifold walls is similar in case of the 1D and the 3D-CFD simulation, the effects on the post-oxidation process might be different. This difference may seem very small at first glance, but once a process has an exponentially self-reinforcing effect, small differences can significantly impact. The results in Figure 4.11 are exemplary for this issue. This is investigated in more detail in chapter 4.3.2.

In addition to this, the necessary computation time for one engine operating point with the usage of the reaction mechanism approach is, compared to usual computation times of a 1D engine simulation, approximately 10 s – 100 s [5], quite high. The computation time of the case displayed in Figure 4.10 is approximately 900 minutes with the usage of a high performance computer (36 CPU nodes and 1.5 TB RAM) (displayed in Figure 4.12). This results from the high amount of reaction equations included species and the amount of subvolumes due to the small discretization length. A direct comparison with the 3D-CFD results needs to be done with caution because the 3D-CFD calculation is parallel made with various cores, while the 1D reaction mechanism approach simulation is made with only one core.

This methodology is the most straightforward method, but as the results show, the direct implementation of the reaction mechanism shows a too high decrease of post-oxidation rate with non-adiabatic walls. The requirements defined at the beginning of chapter 4.3 are not fulfilled with this modelling approach.

The subsequent chapter includes the evaluation of a possible origin for the differences in the post-oxidation rate and the development of a solution approach.

4.3.2 Surrogate Reaction Mechanism Approach

A similar mixing level, the temperature of the exhaust gas and amounts of exhaust emissions combined with an identical reaction mechanism does not deliver a 1D post-oxidation model that fulfills the desired requirements. The differences in 1D and 3D-CFD simulations itself, as well as the characteristic behavior of the reaction mechanism, need to be taken into account in order to develop a post-oxidation model capable to display the effect of changing boundary conditions. The second requirement on the 1D post-oxidation model not fulfilled by the 1D reaction mechanism approach is a short calculation time.

A short calculation time can be achieved by decreasing the amount of necessary equations to be solved for each time step. The discretization length cannot be increased due to its major role in the mixing model. Therefore, the amount of reaction equations and species that need to be taken into account at every time step needs to be decreased.

The omission of reaction equations and species in a reaction mechanism needs to be done with caution. The absence of a supposedly unimportant component might lead to a significant change in mechanism behavior. As discussed in [22] in detail, the identification of controlling reaction equations is not trivial, and even slow reactions could be of great importance. Therefore, the used reaction mechanism is not reduced, but a surrogate mechanism is developed.

Surrogate Mechanism Development

This chapter concentrates on the development of a surrogate reaction mechanism, which is able to display the post-oxidation process in a simplified way with as few reaction equations as possible.
The surrogate mechanism needs to be capable to display the oxidation of CO as well as the oxidation of H_2 (as mentioned in chapter 3.1.2, H_2 is beside its role as a catalyst for the CO oxidation a major energy source of the post-oxidation process).

Chapter 3.2.6 shows that the H_2 oxidation is faster than the CO oxidation inside the exhaust manifold. To display this behavior, the surrogate reaction mechanism needs to include at least two different reaction equations to display these different reaction rates. The chosen reaction equations are:

$$2CO + O_2 = 2CO_2 \tag{4.1}$$

$$2H_2 + O_2 = 2H_2O \tag{4.2}$$

The requirements for these reaction equations is to show a similar reaction rate in comparison with the reaction mechanism used in the 3D-CFD simulations but with modifications in order to prevent an unrealistic behavior on changing boundary conditions as shown in chapter 4.3.1.

The investigation in chapter 3.1 shows that the post-oxidation effect takes place in environments with a temperature above 1000 K. The reaction rates then increase with increasing temperature. This effect is the source of the exponential self-reinforcing behavior of the post-oxidation phenomenon and simultaneously, the source of the flame extinguishing effect for too low temperatures (as mentioned in chapter 3.2.4). Hence, this is the starting point for the surrogate reaction mechanism development.

The subsequent investigation includes a reaction kinetic investigation. The focus of the investigation lies on the temperature span of 1000 K to 1200 K since it is in this temperature range that it is decided whether the post-oxidation will enter a self-reinforced state.

Figure 4.13 shows the mole fraction of CO within a time span of 5 ms. The choice of 5 ms is mentioned in chapter 3.1.1 and is approximately the maximum exhaust gas residence time inside the exhaust manifold. This investigation and the following reaction kinetics investigations are made by usage of Cantera according to the methodology described in chapter 3.1.1 including reaction mechanism 2. It shows the change of the CO mole fraction of a mixture of incomplete burnt exhaust gas in combination with fresh air. The temperature of the 0D reactor is held constant and is varied from 1000 K to 1200 K to a total amount of three simulations. The temperature is held constant, because of the fast temperature increase with increasing oxidation rate. If the oxidation process starts, the temperature reaches the high temperatures within a short time, and the different reaction rates of the post-oxidation process cannot be observed. The pressure is 1 bar.

As mentioned before, the CO mole fraction does not change for a reactor with a constant temperature of 1000 K (at least within 5 ms). With increasing temperature, the mole fraction of residual CO after 5 ms decreases. Besides the CO conversion rate, the ignition delay decreases.

The oxidation rate with a constant temperature decreases with increasing time. This results from the decreasing amount of available CO and O_2 and a constant reaction rate (due to the constant temperature).

The surrogate mechanism needs to be able to display the major effects of the post-oxidation within the time span of 5 ms:

- No CO oxidation for temperatures $\leq$ 1000 K.
- Similar CO mole fraction after a time span of 5 ms for different temperatures.

The ignition delay is – as explained in chapter 3.1 – a result of the radical formation. Since the formation of radicals needs additional reaction equations and species, the ignition delay is not modelled directly with this modelling approach but taken into account indirectly within the development process.

The ratio of remaining CO and H_2 mole fraction after 5 ms regarding the start value of CO and H_2, in this work called Z, is the target value for the surrogate mechanism development.

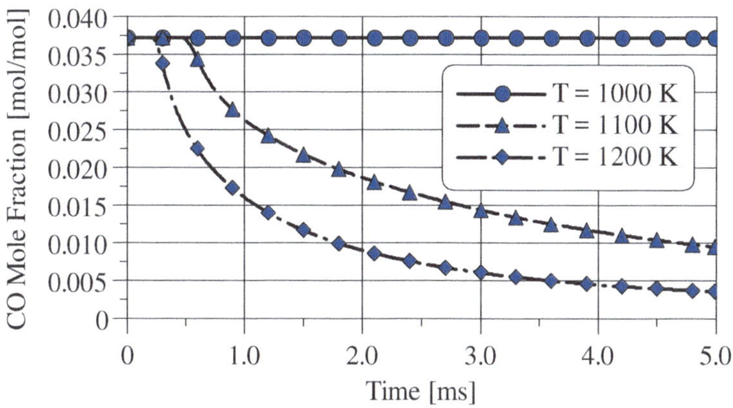

Figure 4.13: *CO* mole fraction decrease due to oxidation with a different constant starting temperature.

$$Z_{CO} = \frac{\chi_{CO_{5ms}}}{\chi_{CO_{start}}} \qquad 4.3$$

$$Z_{H_2} = \frac{\chi_{H_{2_{5ms}}}}{\chi_{H_{2_{start}}}} \qquad 4.4$$

A similar behavior of Z for both reaction mechanism and surrogate mechanism with changing temperature is the key element to be able to display the self-reinforcing effect but also the flame extinguishing effect in a pragmatic approach.

The change of Z_{CO} for a specific temperature range with, besides the temperature, constant starting conditions is displayed in (Figure 4.14). The diagram shows that Z_{CO} is similar to one for temperatures below 1000 K. Then, Z decreases abruptly and gets smaller for increasing temperatures (but with inversely proportional gradient). It is important to stress that this behavior is plausible from the reaction kinetic point of view.

The abrupt and steep increase in reactivity might be the main reason for the implausible behavior of the 1D reaction mechanism approach. If the temperature in the local 1D subvolume is in a temperature range close to 1000 K, a

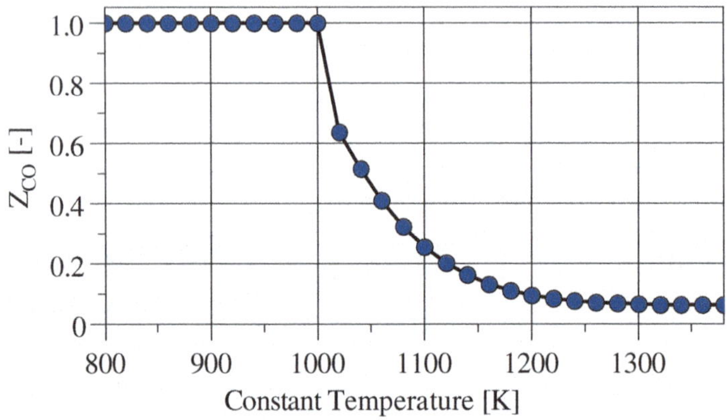

Figure 4.14: Change in unoxidized *CO* after 5 ms with different constant reactor temperatures (reaction mechanism 2).

rather small difference in temperature compared to the 3D-CFD mesh can lead to a major discrepancy.

Especially because of the self-reinforcing behavior of the post-oxidation effect. The previous discussion on the influence of the wall heat flux is a possible source of the deviation in temperature.

In Figure 4.15, the temperature of a local 1D subvolume is evaluated at position 2. It is observable that the temperature is close to 1000 K, as soon as the high temperature exhaust gas – represented by *CO* – and the quite low temperature scavenging air – represented by O_2 – mix up. The temperature range around 1000 K is therefore within the range relevant for post-oxidation. In combination with effects that affect the temperature inside the exhaust manifold – e.g. the heat fluxes at the manifold walls – the temperature inside the exhaust manifold do differ more likely when comparing the 1D and the 3D-CFD simulation results. This increases the probability of a temperature deviation in a temperature range close to 1000 K.

Additionally, the omission of a radial temperature distribution intensifies the problematic effect of the abrupt increasing reactivity. Figure 4.17 shows the pipe cross-section resulting from the 3D-CFD simulation of case #4c at position 2.

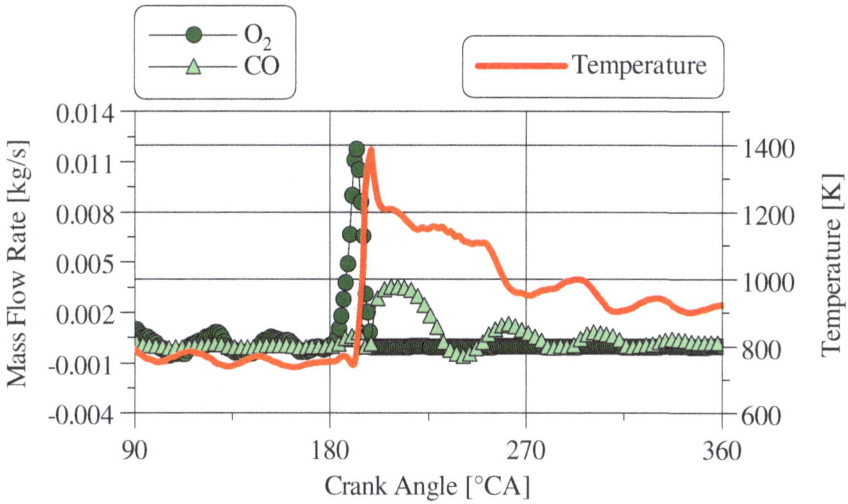

Figure 4.15: Mass flow rate of CO and $O2$ and temperature in position 2 (1D engine model, case #1).

The temperature distribution is evaluated at a time step with a mean temperature at this position of approximately 1000 K. While the mean temperature indicates a specific reaction rate within the 1D simulation domain, the temperature distribution results in several different reaction rates in case of a 3D-CFD simulation. Even if the mean temperature is below 1000 K, the 3D-CFD can show both very low reaction rates and high reaction rates in some regions.

In Figure 4.16, the same position is evaluated at a different time step. The mean temperature is approximately 850 K. The 1D reaction mechanism approach would show a very low reaction rate in terms of CO oxidation. The 3D-CFD simulation, on the other hand, shows in some regions within this surface a quite high reaction rate, since in some areas the temperature is above 1000 K.

An average temperature is a simplification of real temperature distribution. This temperature distribution acts as a smoothing of the reactivity with increasing temperature. In order to be able to represent a realistic behavior of the 1D simulation with respect to the temperature influences, it is important to include

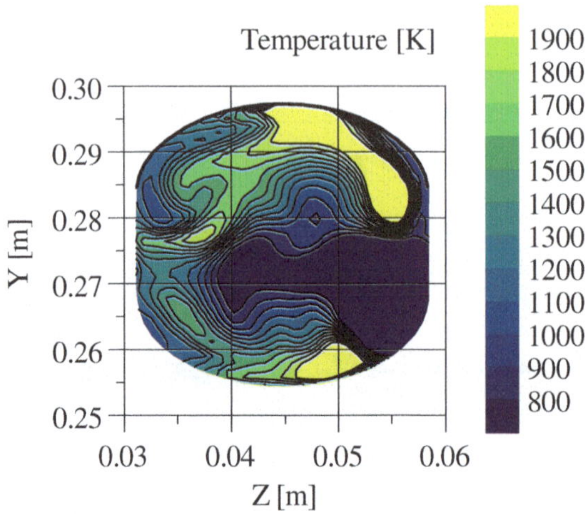

Figure 4.16: 3D-CFD temperature distribution of case #4c in position 2. Surface average temperature = approximately 850 K.

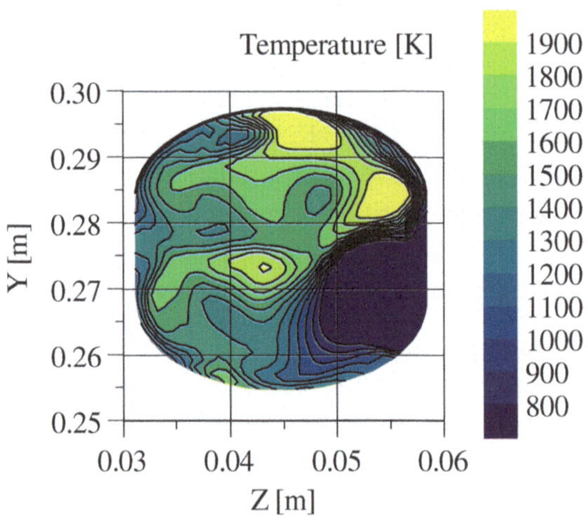

Figure 4.17: 3D-CFD temperature distribution of case #4c in position 2. Surface average temperature = approximately 1000 K.

this behavior in the 1D model. Since the 1D simulation cannot represent a radial temperature distribution, a different approach must be chosen.

The area for temperatures above 1000 K is not necessarily to be modelled with great accuracy. As soon as the exhaust gas and fresh air mixture reach temperatures in the area with a steep increase in reactivity, the released heat will increase the temperature of the mixture. This again leads to an even more increased temperature. The most important effect, which needs to be displayed by the surrogate mechanism, is a steep increase in reactivity but the accuracy in this area is less important as in the temperature range very close to 1000 K.

The Arrhenius parameters of the surrogate mechanism are now set in a way to display the behavior explained above. As discussed in chapter 2.1, the modified Arrhenius equation is defined as shown in equation 4.5.

$$k = AT^b e^{\frac{-E_A}{R_u T}} \quad 4.5$$

The parameters A, b and E_A need to be set.

The abrupt increase in reactivity with increasing temperature requires an exponential growth of the reaction rate constant. This behavior is achieved by choice of a high temperature exponent b. The higher b, the greater the increase with increasing temperature. However, this leads to a very high base level of the reaction rate constant.

To lower this base level, especially for temperatures below 1000 K, a high activation energy E_A can be used. The activation energy describes how much energy is required during a collision of two molecules to start a reaction according to a specific reaction equation. Therefore, a high activation energy leads to a low reaction rate for low temperatures at which the averaged energy of the molecules is limited, and only a few are able to overcome this energy barrier. This behavior is mathematically modelled by a natural exponential function. The increase in temperature leads to an increased gradient of the reaction rate constant but at a very low base level.

A well-thought-out combination of the temperature exponent b and the activation energy E_A leads to a very high increase in reaction rate constant for a temperature close to 1000 K with a quite moderate base level in reactivity. The pre-exponent multiplier A can then tune the overall base level.

The choice of the temperature exponent b and the activation energy E_A is also important to include a smoothing effect into the surrogate mechanism. As discussed earlier, the temperature distribution of a real flow file inside a pipe acts as some kind of smoothing effect on the abrupt increase in reactivity. The choice of b and E_A in a way, the increase in reactivity close to 1000 K is too abrupt might lead to the same issues as discovered with the direct implementation of the reaction mechanism. By doing this, the smoothing due to the temperature distribution can be implemented directly into the oxidation model.

If the increase in reactivity close to 1000 K is similar to the original reaction mechanism, the effect of temperature distribution must be included in a different way. A possible methodology would be determining a Gaussian distribution, based on the pipe cross-section mean temperature. However, since this would lead to an additional modelling step, an increase in calculation time is to be expected. Therefore, the implementation of the temperature distribution is directly integrated into the development of the surrogate mechanism.

Tuning of the model is based on the system reactivity at 1100 K. The choice of 1100 K is characterized by the fact that at this temperature, the changes in reactivity over temperature are no longer as pronounced, in terms of both CO and H_2. With a constant temperature of 1100 K, the reaction mechanism shows for the equation 3.1 of approximately $1.8 \cdot 10^8$ kmol/(m s) for a sample at 1100 K and an ι equal to 0.9. The decrease in CO is, as displayed in chapter 2.1. The decrease in CO of the surrogate mechanism should be approximately similar to this value. The difference in the reaction mechanism and the surrogate mechanism is the responsible oxidizer. While the reaction mechanism is dependent on the OH-radical, the surrogate mechanism relies on the oxidation by O_2. The difference in concentration of these two oxidizers leads to a different reaction rate constant k to have a similar decrease in CO. With a roughly 50 times greater amount of O_2 than OH in its peak concentration, the reaction rate constant of the surrogate mechanism should be approximately 50 times lower than the corresponding value of the reaction mechanism. The very high temperature dependency, the very high activation energy and the tuning of the surrogate mechanism to achieve similar decreases in CO based on this rough analysis on orders of magnitude lead to the Arrhenius equation parameters listed in Table 4.1. The oxidation of H_2 shall show similar behavior with changing temperature, but the whole process is shifted to a higher base level due to the higher reactivity of H_2 and to cover the decrease of H_2 due to the OH-radical formation. The corresponding Arrhenius parameters are also listed in Table 4.1.

Table 4.1: Surrogate mechanism Arrhenius equation parameters.

	A	b	E_A
Equation 4.1	$2.5 \cdot 10^6$	7	$3.7413 \cdot 10^8$
Equation 4.2	$2.5 \cdot 10^7$	7	$3.7413 \cdot 10^8$

In Figure 4.18, the ratio of remaining CO as defined in equation 4.3 is plotted together with the already shown results of the reaction mechanism based simulation. As discussed and therefore targeted, Z_{CO} equals one for low temperatures.

Close to 1000 K, the decrease of Z_{CO} starts for lower temperatures and does not show an abrupt start of decrease. The differences for temperatures higher than 1000 K is, as discussed earlier, not crucial due to the very limited time span, the temperature of a $CO/H_2/O_2$ mixture would remain in this temperature range (because of the very fast heat up of the mixture as soon as the reaction starts). Both curves intersect at approximately 1100 K. This results from the fact that the analysis of the order of magnitude is made close to 1100 K. The difference in Z_{CO} for temperatures above 1100 K is quite small.

A similar investigation is made concerning the H_2 inside the same simulation domain. The even more pronounced abrupt increase in reactivity is displayed in Figure 4.19. The origin of this behavior might be the fact, that the production of the OH-radicals starts with a temperature greater than 1000 K. The oxidation of H_2 also depends on the OH-radical concentration (as discussed in chapter 3.1.2). In combination with a comparatively low quantity of H_2 molecules, the decrease of H_2 with increasing temperature seems unrealistic abrupt but comprehensible. The surrogate mechanism approach does also smooth the abrupt decrease of H_2 in the 1000 K area. The discrepancy of the two curves for temperatures greater than 1000 K is again not crucial, due to the same reasons as explained for the CO oxidation approximation. These results are now compared for different mixture compositions and different constant temperatures. The comparison is made by the difference of the corresponding Z_{CO} of both reaction mechanism and surrogate mechanism. The result is shown in Figure 4.20. The difference in the temperature range above 1000 K and below 1100 K is, as already mentioned,

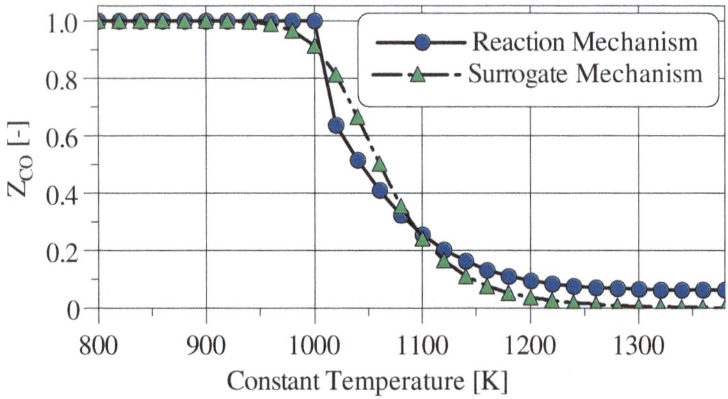

Figure 4.18: Z_{CO} comparison: reaction mechanism used in the 3D-CFD simulation and new developed surrogate mechanism.

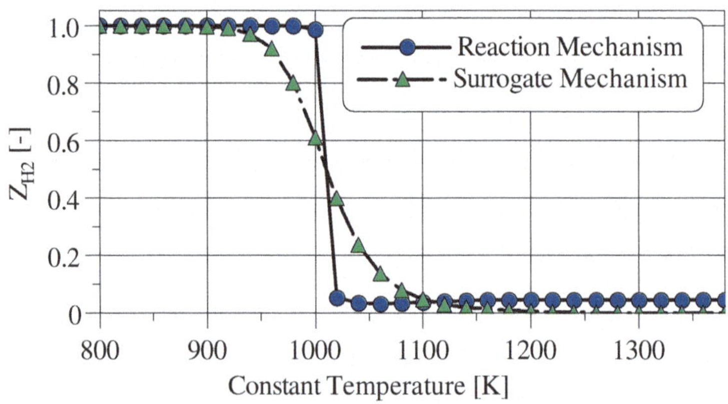

Figure 4.19: Z_{H_2} comparison: reaction mechanism used in the 3D-CFD simulation and new developed surrogate mechanism.

not crucial. The difference between higher temperatures and high ι arises from the way the results are evaluated. The high relative difference is due to the small absolute values of present CO in a mixture with a ι close to 1.

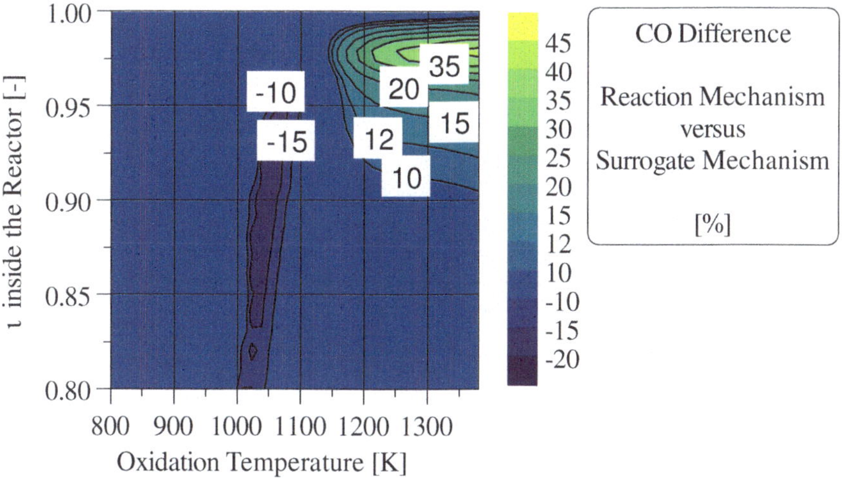

Figure 4.20: $Z_{CO_{\text{reaction mechanism}}} - Z_{CO_{\text{surrogate mechanism}}}$ with different constant temperatures and different species compositions.

The differences in case of the H_2 oxidation approximation are displayed in Figure 4.21. The areas with increased differences are similar to the evaluation concerning CO. The relative discrepancy close to 1000 K is higher, since the abrupt behavior of the H_2 decrease due to oxidation is more pronounced in comparison to CO. Nevertheless, the absolute quantities of H_2 inside the investigated reaction kinetic system is quite small and therefore, the absolute discrepancy is small. The surrogate mechanism shows a good approximation of the reaction mechanisms behavior. The temperature range close to 1000 K is smoothed, and the steep increase in reactivity for temperatures higher than 1000 K is also realized.

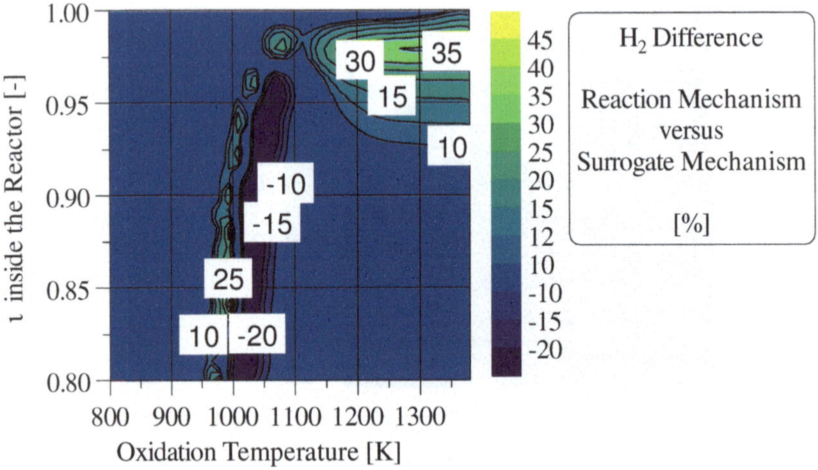

Figure 4.21: $Z_{H_{2}\text{reaction mechanism}} - Z_{H_{2}\text{surrogate mechanism}}$ with different constant temperatures and different species compositions.

Implementation

Implementing the new developed surrogate mechanism is made by using a chemical reaction template called "GlobalReaction"-Template. The template requires the parameters of the Arrhenius equation for the implementation of the surrogate mechanism. The parameters need to be implemented in the template with the units cm and mole. The parameters in Table 4.1 are defined in meter, kmol. Since the conversion of the units is not completely trivial, it is shown in the following.

The unit of the reaction rate constant changes with changing reaction order. As displayed in chapter 2.1, the reaction order can be found within the exponents of the species concentrations. Since the unit of the equations left side needs to be related to the change in concentration for a specific time step, the reaction rate constants dimension changes with reaction order. This needs to be considered with the unit conversion.

$$k[\text{mol, cm, s}] = (\frac{(0.01)^3}{0.001})^{m+n-1}[\text{kmol, m, s}] \quad 4.6$$

$m = \quad 1$

$n = \quad 1$

The surrogate mechanism is despite the good approximation of the results shown before not capable of displaying the chemical ignition delay due to the formation of radicals inside the exhaust manifold. This chemical ignition delay leads to a temporally offset in post-oxidation for every exhaust stroke. To mimic this behavior by the usage of the available tool-set, the chemical reaction template and therefore, the implemented surrogate mechanism is not applied to the valve area. Therefore, the reaction can start at the earliest in the adaptor area. This setup is shown schematically in Figure 4.22.

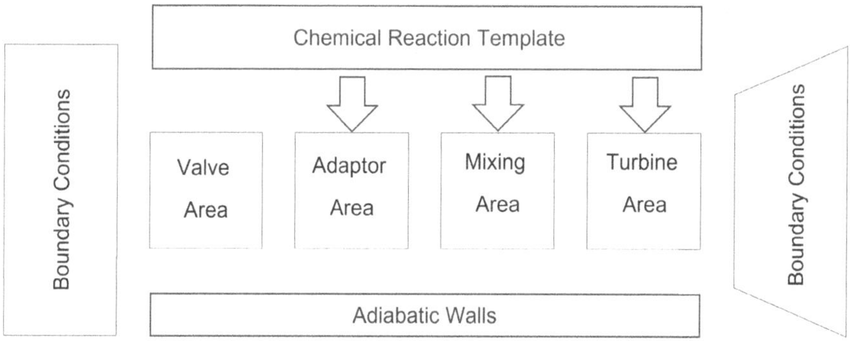

Figure 4.22: Schematic illustration of the 1D engine model with applied chemical reaction template, including the surrogate mechanism.

Discussion and Results

The 3D-CFD simulation results and the results of the 1D engine simulation inclusive implemented surrogate mechanism are shown in Figure 4.23. The chosen engine operating point is case #4. The post-oxidation rate of the 1D results is higher than the corresponding post-oxidation rate of the 3D-CFD simulation. This might be because of the higher reactivity of the surrogate mechanism in areas with a temperature of approximately 950 K to 1000 K and the higher reaction rate for areas with a temperature above 1100 K. Besides, the workaround to mimic the chemical ignition delay is of course not ideal and therefore subject of errors.

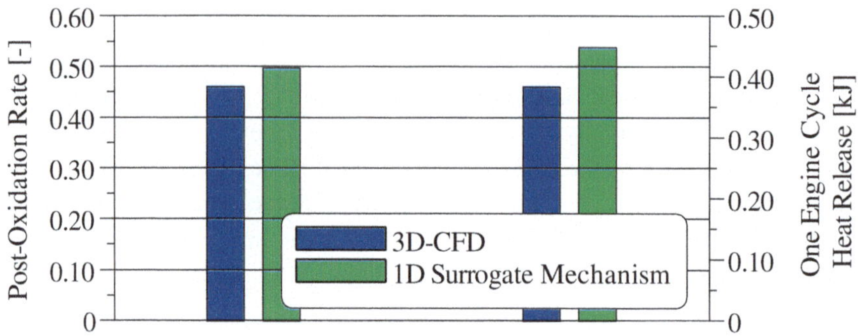

Figure 4.23: Difference in post-oxidation rate and chemical heat release of the 3D-CFD simulations and the 1D model approach including the surrogate mechanism.

The variation of the boundary conditions in terms of non-adiabatic exhaust manifold walls, as already done and discussed for the 3D-CFD simulation and the 1D reaction mechanism approach, is shown in Figure 4.24 for the surrogate mechanism approach. The post-oxidation rate also decreases due to the heat flux at the exhaust manifold walls. Compared to the 3D-CFD simulation, the decrease is still overestimated by the 1D engine approach but to a lesser extent as with the 1D reaction mechanism approach. This is the effect of the compensation measures made in the oxidation model development – the smoothing of the reactivity temperature dependence. The overrating of the wall heat losses due to the missing radial temperature gradient is still present due to its origins in the 1D domain itself, but the accompanying effect of inhibiting self-reinforcing effects is limited. The self-reinforcing effects start with a lower temperature – resulting in an overrated post-oxidation process in case of adiabatic walls – but partially compensates the issue of missing radial temperature gradient.

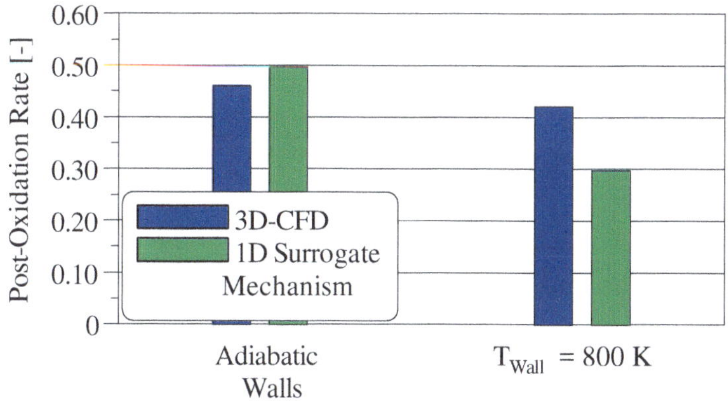

Figure 4.24: Difference in post-oxidation rate of the 3D-CFD simulations and the 1D model approach including the surrogate mechanism concerning changing wall boundary conditions.

The reduction of the necessary reaction equations and therefore the corresponding species decrease the necessary computation time for each simulated engine operating point. An additional reduction of the computation time is achieved by omitting chemical connection to the valve area. The resulting computation time compared to the 3D-CFD simulation and the 1D reaction mechanism approach is displayed in Figure 4.25. The compared engine operating point is case #4.

The 1D surrogate mechanism approach shows a significantly lower computation time. While the 1D reaction mechanism approach needs approximately 900 minutes, the new approach needs approximately 22 minutes.

In the following, the 1D surrogate mechanism approach is subjected to further sensitivity analysis and if available in combination with validation using the 3D-CFD simulation results.

In Figure 4.26, the engine operating points boundary conditions are changed in the way described in Table 3.5. The used engine operating points are case #4c, case #4c1 and case #4c2. The wall heat losses at the exhaust manifolds walls are modelled again with a constant wall temperature of 800 K. The models'

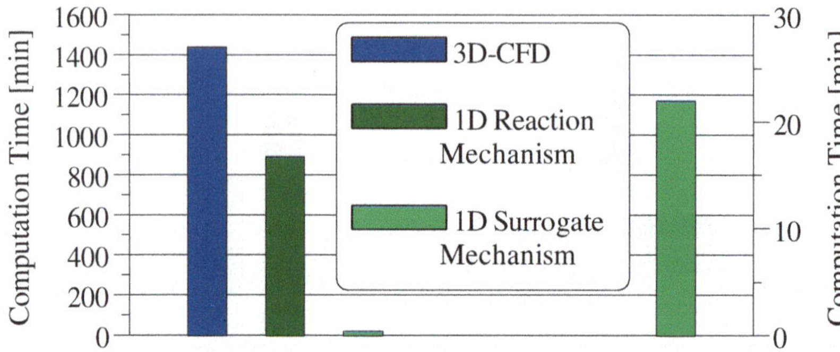

Figure 4.25: Comparison of computation time of case #4: 3D-CFD simulation, 1D reaction mechanism simulation and 1D surrogate mechanism simulation.

sensitivity on changing exhaust gas temperatures is tested with a change on the exhaust gas peak temperature as described in chapter 3.2.2.

The 1D post-oxidation model is capable of displaying the decrease of the post-oxidation rate with decreasing exhaust gas peak temperature. With a decrease in the post-oxidation rate for an exhaust gas peak temperature decrease of 100 K, the 1D post oxidation model shows a decrease of 18 %, while the 3D-CFD simulation shows a decrease of 9 %. With a further decreasing exhaust gas peak temperature, the 1D post-oxidation model shows a decrease of 9 % while the 3D-CFD simulations post-oxidation rate decreases by 23 %.

The 1D post-oxidation model shows a weakening of the self-sustaining effect for a less decreased exhaust gas peak temperature in comparison to the 3D-CFD results. The reason for this behavior might be the differences in the order of discretization length and the differences of the simulation domains itself. While a coarser discretization length in case of the 1D simulation results in lower peak temperatures, a lower exhaust gas peak temperature has a greater influence on the post-oxidation rate. The areas with a very high temperature are the source of the self-sustaining effect of the post-oxidation process. The reaction speed of the incomplete burnt exhaust gas mixture increases exponential with increasing temperature. As long as the averaged peak temperatures of the 1D post-oxidation model are high enough to start a fast post-oxidation process, the differences of the 1D and the 3D model are limited (Figure 4.24). A decrease of the exhaust gas peak temperature can result in average temperatures inside the discretized pipes, which is partially not able to start a post-oxidation process. Beside the lower reaction rate, the effect of a less strong self-sustaining effect results from this temperature decrease. This results in a greater decrease in the post-oxidation rate as expected by comparing to the 3D-CFD results. A similar decrease in the post-oxidation rate of the 3D-CFD simulation can be seen for a greater decrease in exhaust gas peak temperature. With a finer discretization, the combined effect of a lower reaction rate and a less strong self-sustaining effect is reached with a lower exhaust gas peak temperature.

An additional effect could be, as stated earlier, the overrated amount of heat flux at the exhaust manifold walls. Therefore, a source of difference between the two simulation results (3D-CFD simulation and 1D simulation) is located in the simulation domain itself.

Nevertheless, the 1D post-oxidation model shows a comprehensive behavior on the exhaust gas peak temperature variation.

The following sensitivity analysis is based on a geometric change of the exhaust manifold. The adaptor, which is necessary for the test-bench investigation, is removed, and the resulting post-oxidation rate is evaluated. In Figure 4.27, the results of this investigation are shown. The omission of the adaptor reduces the post-oxidation rate by approximately 2 %. This small influence can be explained with the rather low mixing level in this area combined with a small residence time span of the exhaust gases. The bulk of the already mixed fluid is then oxidized in the mixing zone instead.

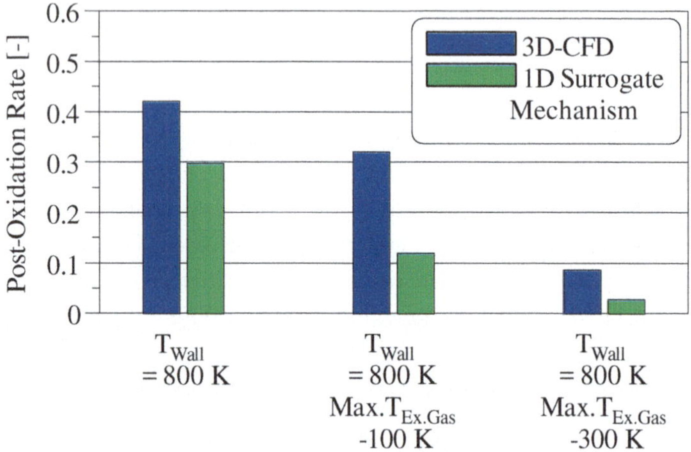

Figure 4.26: Post-oxidation rate comparison of the sensitivity analysis simulation result of the 3D-CFD simulation and 1D simulation inclusive surrogate mechanism approach.

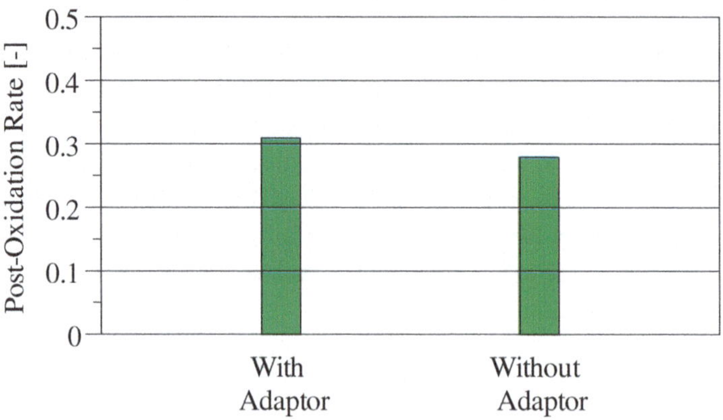

Figure 4.27: Post-oxidation rate of case #4c with and without adaptors.

4.4 1D Post-Oxidation Model Validation

The 1D-post oxidation model can now be applied to the three remaining engine operating points including a reaction mechanism and adiabatic exhaust manifold walls, listed in Table 3.4 (case #1, case #2 and case #3). The used 1D engine model does not contain only the exhaust manifold but also the combustion, the turbocharger and the intake manifold. Therefore, the activated post-oxidation process shifts the engine operating point slightly due to changing exhaust gas enthalpy and a resulting different turbocharger operating point. However, the difference is rather small, and the results can still be compared.

Figure 4.28 shows the resulting post-oxidation rates for all four engine operating points. The post-oxidation rate is quite similar for all investigated operating points (approximately 46 %) – except case #1 – but the corresponding heat release inside the exhaust manifold differs (Figure 4.29). The post-oxidation model is capable of displaying the post-oxidation rates as well as the corresponding heat release. The increase in the post-oxidation rate of case #1 compared to the remaining cases can be reproduced but in a restrained way. The 1D post-oxidation model overestimates the post-oxidation rate for case #2, #3 and #4 by up to 5 %, while the post-oxidation rate of case #1 is underestimated by 5 %.

This comparison shows, that the post-oxidation model is capable to display the heat release due to post-oxidation inside the exhaust manifold for all four available engine operating points.

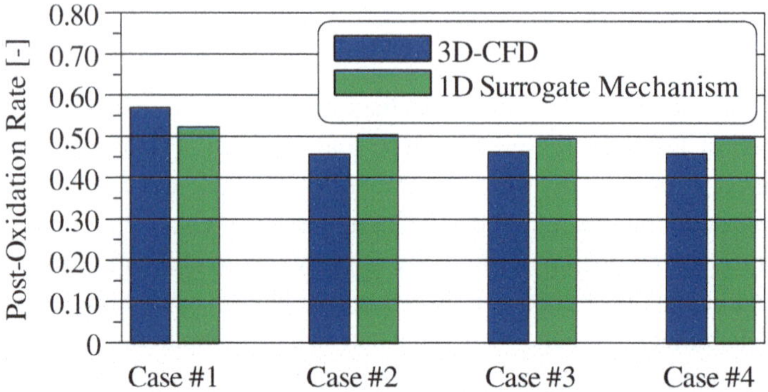

Figure 4.28: Post-oxidation rate comparison of the 3D-CFD simulations and the 1D surrogate mechanism simulations including the whole 1D engine model.

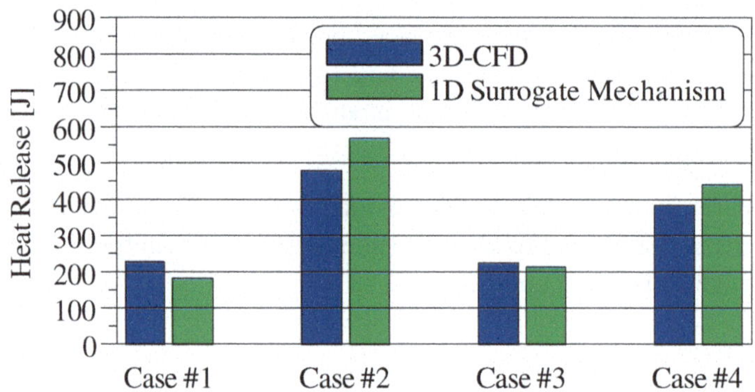

Figure 4.29: Heat release comparison of the 3D-CFD simulations and the 1D surrogate mechanism simulations including the whole 1D engine model.

The defined requirements on the 1D post-oxidation model do also include a similar heat release distribution inside the exhaust manifold. As discussed earlier, the exhaust manifold is subdivided into four subareas. The volume of each subvolume in both simulation domains is similar. The implementation of the oxidation model includes the omission of chemical reactions at the valve area. The resulting heat release distribution for case #1, #2, #3 and #4 is illustrated in Figure 4.30. For reasons of a clear presentation, the exhaust manifold areas numbered: valve area = 1, adaptor area = 2, mixing area = 3 and turbine area = 4.

A comparison of the 3D-CFD results and the 1D surrogate mechanism model results show a similar characteristic in all four evaluated cases. The absence of heat release in the 1D valve area is as already mentioned a direct consequence of the implementation. The heat release of the 1D model seems instead been relocated into the adaptor area. Since the mixture that has not been oxidized inside the valve area must necessarily be of high reactivity, this behavior is comprehensible. A second characteristic of the simulation domain comparison is the lower heat release proportion of the 1D model in the turbine area. While the heat release inside the mixing zone is in all four cases higher with the usage of the 1D model, the heat release inside the turbine area is always lower. This might be either a direct consequence of a different flow behavior within the 3D-CFD and 1D approach or the reactivity of the 1D model in the adaptor and mixing area is overestimated. Therefore, the leftover mixed fluid is lower as in the corresponding 3D-CFD simulation.

Nevertheless, the 1D post-oxidation model approximates the heat release distribution very well. The rather low heat release in the adaptor and turbine area, as well as the high heat release share in the mixing area.

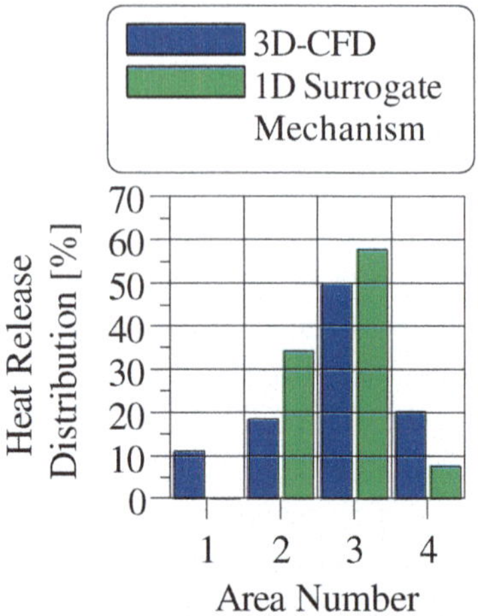

(a) Case #1

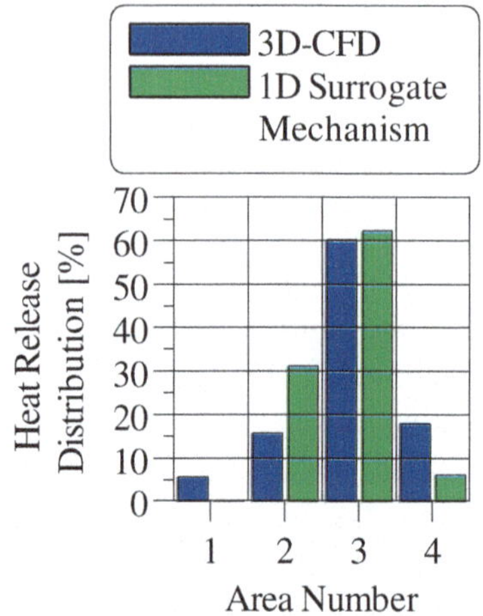

(b) Case #2

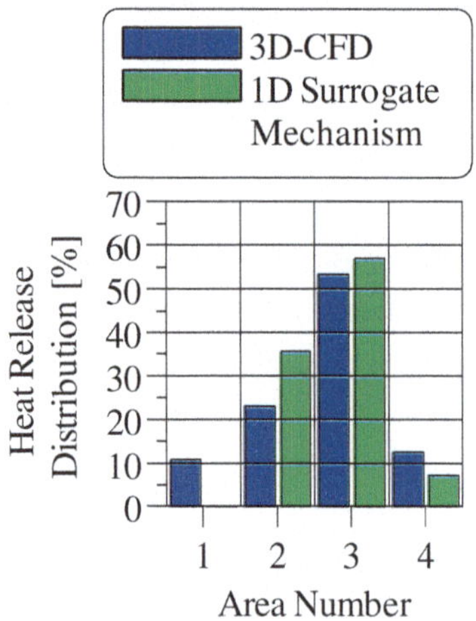

(c) Case #3

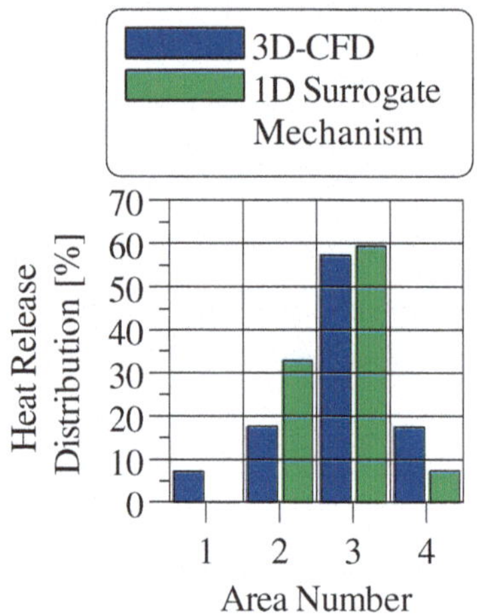

(d) Case #4

Figure 4.30: Heat release distribution inside the exhaust manifold comparison of the 3D-CFD simulation results and 1D simulation results including the surrogate mechanism.

4.5 1D Post-Oxidation Model Discussion

This chapter is dedicated to the discussion of the 1D post-oxidation model. The advantages are highlighted, but the focus of this chapter is the limitations of the approach. An emphasis on the model approach restrictions is important since the outcome of this development is a 1D simulation approach of a problem definition, which is mainly controlled by three-dimensional effects.

The main advantage of the 1D model approach is clearly the decrease in computation time. The 3D-CFD simulations are carried out on a high performance computer (36 CPU nodes and 1.5 TB RAM). The 1D model approach is calculated on the same high performance computer but with only one node instead of several cores. Still, the computation time of the 1D model is about 63 times lower than the 3D-CFD simulation.

The model is able to reproduce the heat release inside the exhaust manifold in each evaluated 3D-CFD case. The heat release distribution inside the exhaust manifold is satisfying, although the heat release inside the valve area is redistributed – as compensation for the omission of the chemical ignition delay. The heat release inside the turbine area is underestimated for each evaluated case. This might have a slight influence on the temperature of the fluid at the turbine but since the overall heat release in similar and the heat release in the mixing area is even greater, this difference is most likely negligible.

The differences in heat release distribution might result from the simplifications made during the mixing model development, more precisely in the evaluation methodology of the mixing inside the exhaust manifold. The mixing level evaluation made in chapter 3.2.6 only considers the axial mixing inside the exhaust manifold. The radial mixture is averaged over the entire evaluation surface and therefore declared as mixed. It is, of course, to be assumed that the radial mixing is not homogeneous. Therefore, the mixing level has – by definition – an overestimating tendency regarding the actual mixing at each position. The tuning of the 1D model considers the axial mixing only. This leads to a possible overestimation of the mixing in the 1D model approach. A direct consequence of this simplification-based overestimation might be the tendency of the 1D post-oxidation model to overestimate the post-oxidation rate as displayed in Figure 4.28. An overestimated mixing leads to an overestimation in heat release due to post-oxidation. The difference between three-dimensional mixing and axial mixing might also be a source of the differences in heat release distribution (Figure 4.30). In areas with rather limited radial mixing (e.g. the adaptor area), the mixing might be overestimated. This might lead to a shift in heat release distribution.

The incapability of the 1D domain to display a radial spatial distribution might be the origin of another issue of the post-oxidation model. The absence of a radial discretization has the consequence that not only the mixing but also temperature of the fluid is averaged over the entire pipe cross-section. This manifests itself as follows:

First, the average temperature may calculate a different wall heat flux for the same constant wall temperature. This results from the temperature difference between fluid and wall. A non-homogeneous distribution of the temperature results in both, a higher and a lower temperature compared to the average temperature, as shown in Figure 4.17. Within the 3D-CFD simulation, this causes – depending on the flow field inside the pipe – both hot and cold fluid close to the wall. Therefore, an identical wall temperature in the 1D model does not necessarily result in the same heat flux at the exhaust manifold walls. Due to the dependence of the temperature distribution within a 3D-CFD pipe cross-section on the actual flow field, a corresponding modification of the 1D wall temperature is not purposeful.

Second, a radial temperature distribution shows a direct impact on the reaction kinetics. Beside the just discussed difficulties regarding wall heat fluxes, the radial temperature distribution has a smoothing effect on the temperature dependency of the reaction rates. The 1D model approach delivers for every pipe cross-section an averaged scalar temperature. This results in an averaged scalar reaction rate. The radial discretization of the 3D-CFD approach shows, in contrast, a reaction rate for each dissolved scalar temperature within the temperature distribution. The result is a reaction rate distribution whose averaged value is not necessarily similar to the averaged scalar reaction rate of the 1D simulation approach (due to the exponential proportional temperature dependency of the reaction rate). This behavior needs to be implemented in order to achieve a realistic post-oxidation model behavior. A possible approach is the evaluation of a Gaussian distribution of the scalar temperature at each pipe cross-section within the 1D model. This Gaussian temperature distribution would also lead to a Gaussian reaction rate distribution and therefore, a similar smoothing effect as in case of the 3D-CFD simulation. Since the reaction kinetics simulation needs to be made with each reaction rate of this Gaussian distribution, an enormous increase in computation time would be the result. With a short computation time as a pre-defined requirement of the post-oxidation model development, a different approach is chosen.

The influence of the wall heat fluxes, as well as the influence of temperature distribution, is a well-known phenomenon in reaction kinetics related model development. An example is given in [16], [17]. The modelling of nitrogen oxides during the combustion process is also affected by wall heat fluxes and resulting temperature distribution. The reaction rates temperature dependency of the nitrogen oxides formation also shows an abrupt behavior at a specific temperature limit. A common approach is the definition of different temperature zones to take the temperature distribution into account. Since the entire exhaust manifold is taken into account in this work, this approach is, in consideration of the increase in calculation time, not purposeful.

The chosen methodology allows a smoothing of the abrupt reaction kinetics behavior while keeping the computation time as low as possible.

The post-oxidation model results, as displayed in Figure 4.26, show a permanent underestimation of the chemical heat release inside the exhaust manifold by the post-oxidation model. This suggests the idea that model tuning improves the overall result of post-oxidation estimation. However, since the two submodels are each based on theoretical principles, no specific tuning parameter has been integrated into the model development. Furthermore, overall model tuning at a specific operating point is critical. The entire subject of post-oxidation is characterized by self-reinforcing effects and, concerning reactivity, by steep gradients. Adjusting the post-oxidation model to more accurately represent a particular specific effect, such as the dependence of post-oxidation on exhaust gas peak temperature, does not necessarily lead to a more realistic overall result. Therefore, model tuning is deliberately omitted in this work.

5 Influence of Post-Oxidation Effects on Engine Behavior

The previous chapter is dedicated to the comparison of 1D post-oxidation simulation results and 3D-CFD post-oxidation simulation results. Since the 1D post-oxidation model is, based on the validations made, considered capable of estimating the heat release, additional engine operating points can now be considered. This application of the 1D post-oxidation model must always be accompanied by consideration of the weaknesses of the 1D approach. Although the mode approach shows a very good overall behavior concerning absolute accuracy as well as concerning boundary condition based tendencies, the results must be evaluated with the necessary caution.

The subsequent chapter deals with an investigation of the post-oxidation in a wider range within the engine map. By usage of the new developed 1D post-oxidation model, this investigation can include various engine operating points with a manageable computation time.

5.1 Low-End Torque Investigations

The post-oxidation model is now applied to a 1D engine model, which corresponds to the test-bench engine (as briefly discussed in chapter 3 and detailed explained in [1]) but without the changes made to apply the measurement applications. These changes include the already discussed changes in exhaust manifold geometry (adaptors), changes in the engine operating strategy (valve timings, the start of injection and others) and the exchange of the turbocharger system.

In order to implement the post-oxidation model, the exhaust manifold is modelled as described in chapter 4.2. This includes a fine spatial discretization as well as the rearrangement of the exhaust manifolds geometry in order to

J. Przewlocki, *Simulative Investigation of Post-Oxidation in the Exhaust Manifold of SI Engines*, Wissenschaftliche Reihe Fahrzeugtechnik Universität Stuttgart,
https://doi.org/10.1007/978-3-658-36378-9_5

fit the actual geometry of the exhaust manifold, of course at the expense of calculation time.

In the following, the mentioned 1D engine model is modified in terms of valve timing. It is important to know that this investigation aims to investigate the behavior of the post-oxidation model. The modified valve timings, which are implemented into the engine model are, therefore only coarse estimations and not well-tuned and optimized values. This results in a 1D engine model, which fits the requirements of this investigation and is able to show basic tendencies in performance changes. It is not claimed that these changes do fit the requirements of a realistic and well-tuned engine application. Therefore, the absolute values in, for example, fuel consumption need to be treated with care. In addition to this, the exhaust manifold modification can influence the behavior of pressure pulsation inside the exhaust manifold. Therefore, the implemented valve timings for some engine operating points might not be optimal anymore. Since this investigation focuses on the low-end torque inclusive scavenging, the valve timings which are not inside this area are not changed. This might again lead to unrealistic results in other parts of the engine map. These uncertainties do not affect the outcome of this investigation and are, therefore, not considered further.

In Figure 5.1, the circular markers show the full load of the engine for an engine speed variation. These simulations are performed with a deactivated post-oxidation model in terms of chemical heat release. The maximum torque can be held for an engine speed of 1400 rpm and higher. An engine speed below 1400 rpm shows a decrease of torque due to decreased mass flow rate. This reduced mass flow is not able to drive the turbine and therefore, the compressor to an extent, which is high enough to increase the intake mass flow. This mass flow rate can be increased by realizing scavenging by the usage of valve overlap. This increase in mass flow is made for engine operating points between 1400 rpm and 2000 rpm. A further increase of engine speed range with a maximum torque can be achieved by an extent of the scavenging engine operating points. The engine is still operated with an overall stoichiometric air-fuel ratio. This can be achieved with rich in-cylinder combustion for engine operating points with applied scavenging valve timings.

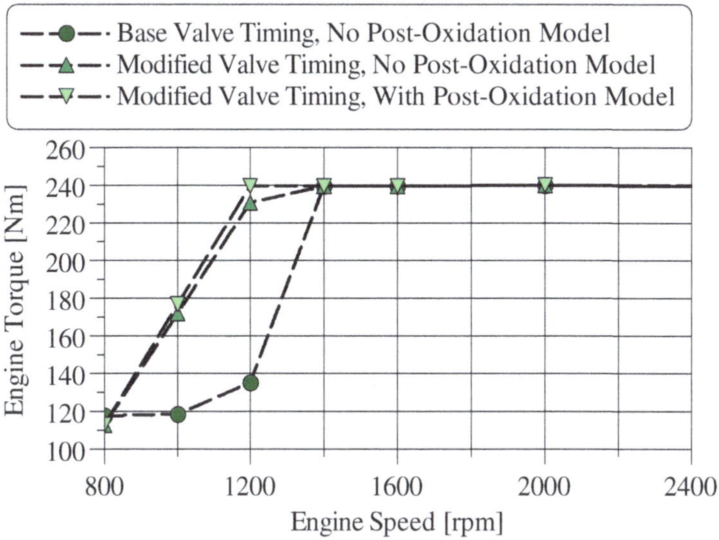

Figure 5.1: Change of the maximum engine torque due to additional scavenging and additional scavenging including the consideration of the increased enthalpy by the 1D post-oxidation model.

The change in valve timing to increase the scavenging air mass flow for all engine operating points with an engine speed below 2000 rpm and boost pressure above 1.4 bar leads to a shift of the low-end torque to a lower engine speed. The low-end torque is now located slightly above 1200 rpm. These results are shown in Figure 5.1 and marked with an upward triangle. The used engine model does again not include the heat release due to post-oxidation effects inside the exhaust manifold. Therefore, the enthalpy of the exhaust gas is underestimated. The activation of the chemical heat release predicted by the post-oxidation model allows an investigation on the actual shift in low-end torque with applied scavenging valve timings.

In Figure 5.1, the downward triangle marked results shows the maximum torque, including scavenging valve timings and activated post-oxidation model. The low-end torque shifts again towards a lower engine speed (down to 1200 rpm). The increase in exhaust gas enthalpy increases the turbine power to the extent that the waste-gate needs to be opened to maintain a maximum torque of 240 Nm.

Figure 5.2 shows the post-oxidation rate and the corresponding chemical heat release for the full load engine points close to the low-end torque. At 1200 rpm, the heat release is quite high, which is the reason for the increase in maximum engine torque at this engine speed. Note that the increase could have been higher without the active turbine waste-gate torque limiting. The limited chemical heat release for 1000 rpm inhibits an effective increase in engine torque (Figure 5.1, upward and downward triangles).

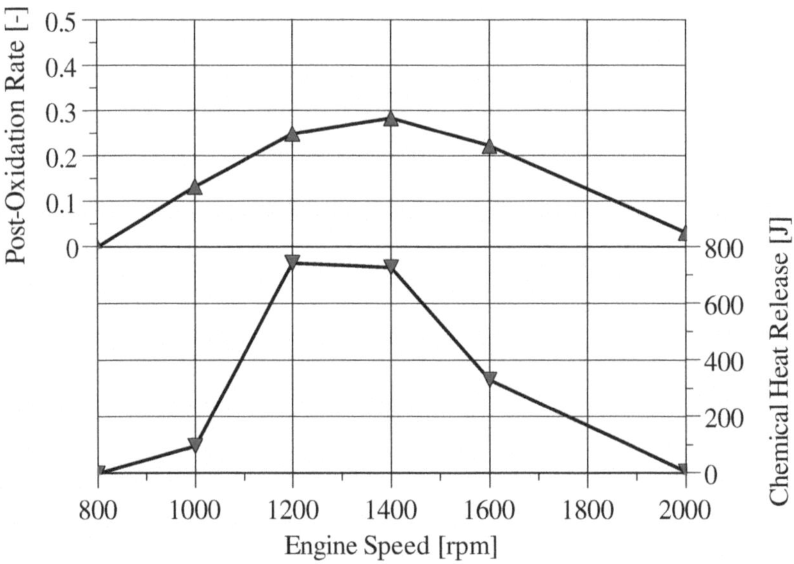

Figure 5.2: Post-oxidation rate and heat release of the test-bench engine with original application settings (as discussed above).

Without available 1D post-oxidation model, the underestimation of the exhaust gas enthalpy needs to be modelled in a different way in order to match, for example, test bench results. This may lead to similar results but also results in less well-tuned 1D engine models or questionable simplifications.

A possible workaround to model a similar increase in boost pressure without a post-oxidation model in comparison to a 1D engine model with included post-oxidation model is the modification of the turbine efficiency. By increasing the turbine efficiency, the turbine can generate more power out of a constant mass flow with a constant enthalpy. In Figure 5.3, the current waste-gate diameter of the turbine out of the model with the implemented post-oxidation model is also used in a 1D engine model without the post-oxidation model. Then, the turbine efficiency is increased by using a multiplier to generate the same boost pressure and therefore, the same engine torque. It can be observed that a turbine efficiency increase of 13 % is necessary to display the same engine torque (at least for this example). This means that current 1D engine models with scavenging engine operating points at the low-end torque might have an overrated and unrealistic high turbine efficiency at these engine operating points.

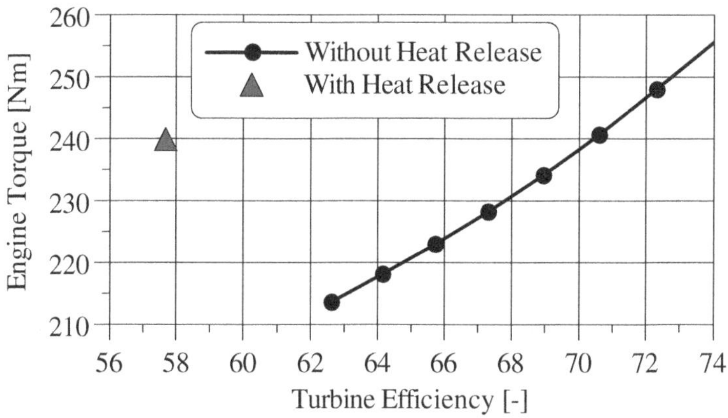

Figure 5.3: Increase in engine torque by modifying the turbine efficiency multiplier.

A second methodology to increase the exhaust gas enthalpy is modifying the heat transfer coefficient at the exhaust manifold pipe walls. In Figure 5.4, the external heat transfer coefficient of the pipes very close to the exhaust valve is varied. The higher the external heat transfer coefficient, the colder the exhaust manifold walls. A lower wall temperature leads to a lower exhaust gas enthalpy and therefore to lower engine torque.

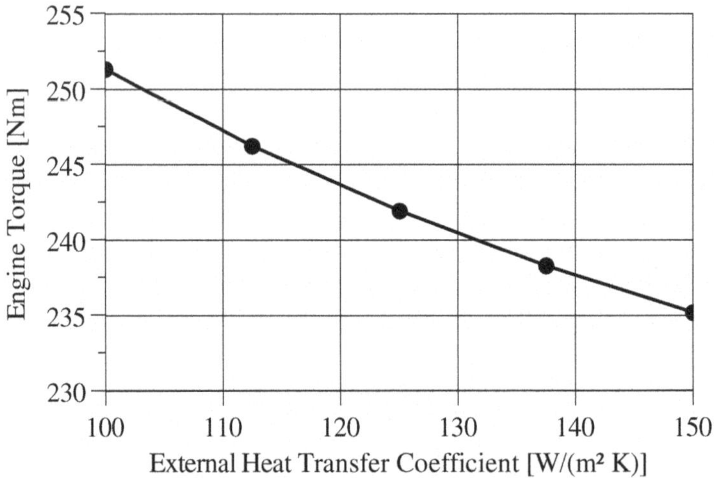

Figure 5.4: Increase in engine torque by modifying external heat transfer coefficient.

Both methodologies can reproduce the results of the 1D engine model with implemented post-oxidation model. Hence, these model modifications cannot directly react to changing boundary conditions, which are related to the post-oxidation rate, these modifications can result in an overestimation or underestimation of the maximum engine torque at low engine speed. Well-tuned models would need either a very high amount of test bench results at areas inside the engine map with active scavenging or the results of a 3D-CFD investigation. Both ways are very elaborate.

5.2 Engine Map Investigation

The application of the 1D post-oxidation model allows the investigation of the post-oxidation in the entire engine map. Figure 5.5 illustrates the resulting heat release inside the exhaust manifold due to post-oxidation. Figure 5.6 shows the corresponding trapping efficiency, and Figure 5.7, the post-oxidation rate of the engine operating points. It is clearly observable that the heat release is proportional to the trapping efficiency. This correlation is because of the amount of incomplete burnt exhaust gas that enters the exhaust manifold. Lower trapping efficiency represents a higher share of scavenging air and, therefore, represents a lower cylinder air-fuel ratio. The trapping efficiency results from the valve timing, the available boost pressure and the exhaust pressure.

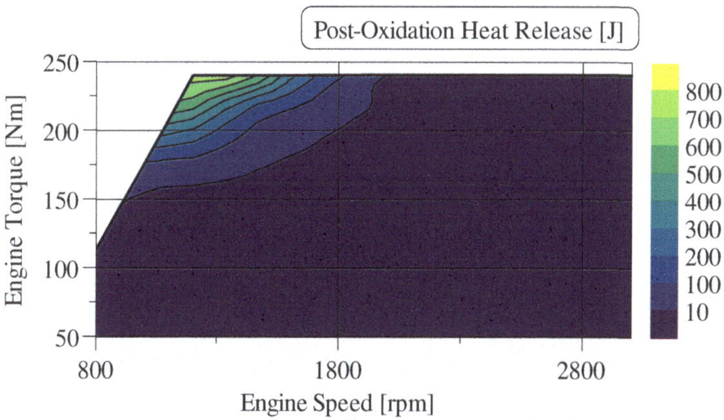

Figure 5.5: Heat release due to post-oxidation inside the exhaust manifold with changing engine speed and engine torque according to the post-oxidation model.

The post-oxidation rate, on the other hand, is not directly proportional to the trapping efficiency. In this case, the maximum can be found at engine operating points with a higher engine speed but with a still relatively low trapping efficiency. This result is quite comprehensible. A changing engine speed leads to changing cylinder-to-cylinder interaction. This affects the mixing behavior. A connection to the trapping efficiency is also observable. This can be

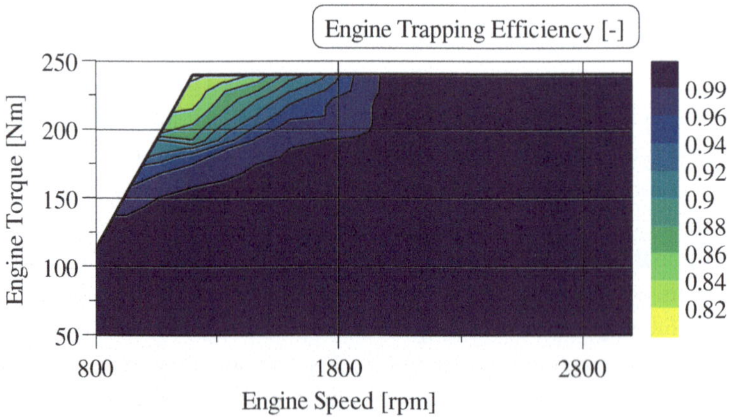

Figure 5.6: Trapping efficiency of the engine due to the scavenging engine operating strategy.

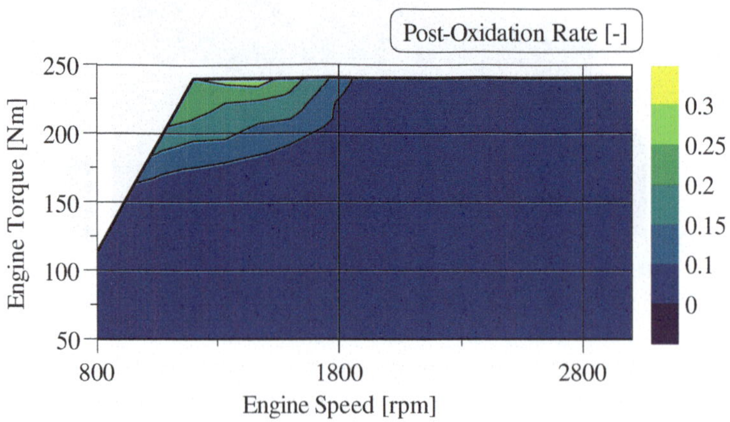

Figure 5.7: Post-oxidation rate within an engine speed and engine torque span according to the 1D post-oxidation model.

explained with the increasing amount of releasable energy entering the exhaust manifold with decreasing air-fuel ratio inside the cylinders. A too low amount of releasable heat lowers the self-sustaining effects of the post-oxidation effect. Because of this, the post-oxidation rate decreases due to lower reaction rates.

6 Conclusion and Outlook

The 1D post-oxidation model behaves in a comprehensible way inside the entire engine map. The model might be unable to display three-dimensional effects like a complex vortex generation and has only a simplified spatial temperature and species distribution. Moreover, the surrogate mechanism does not take into account the generation of radicals. Nevertheless, the most relevant effects are evaluated and successfully implemented into the 1D post-oxidation model by the usage of alternative methods that can be applied to the 1D simulation domain. The generation of vortices in front of the turbine volute, for example, is a highly three-dimensional effect depending on the interaction of turbulence, flow separation and more. Since this effect is observable for each investigated 3D-CFD simulation, it is possible to replace the mixing effect of a vortex by the usage of the mixing increasing effect of a high discretization length within the 1D engine model. This methodology might be unable to completely describe the mixing behavior inside the exhaust manifold in great detail and most certainly fails to describe some relevant effects that result from a very specific combination of boundary conditions.

However, and this is the most important ability of the 1D post-oxidation model, it is able to display the major effects of the interaction of mixing and reaction rate.

The evaluation of the results needs always be made with the background information on the abilities of the model:

- A generally logical mixing behavior along the exhaust manifold flow path.
- A comprehensible dependency of the reaction rates with changing temperature.
- A transformation of the exhaust gas species.
- A conclusive increase in enthalpy due to the species transformation.

In this work, it is refrained from tuning the model additionally by the usage of a tuning parameter. A tuning parameter in terms of increasing or decreasing

J. Przewlocki, *Simulative Investigation of Post-Oxidation in the Exhaust Manifold of SI Engines*, Wissenschaftliche Reihe Fahrzeugtechnik Universität Stuttgart,
https://doi.org/10.1007/978-3-658-36378-9_6

reactivity might lead to a better result of single engine operating points in comparison to the 3D-CFD results. As a result, the overall behavior of the model might also be affected. The post-oxidation process is a highly self-reinforcing phenomenon, and the 1D post-oxidation model reflects this. These tendencies of the model are a result of the simplifications of the 1D simulation domain. Coarser spatial distribution and even missing radial spatial distribution lead to an increased averaging of temperature. In combination with the highly temperature dependent oxidation submodel that itself contributes to the gas temperature inside the exhaust manifold, a slightly different temperature might lead to significantly different results.

The heat flux of the exhaust manifold walls does additionally add an uncertainty factor to the post-oxidation model. The 1D modelling of the wall heat transfer is a complex field of research and a well-known challenge in the 0D/1D modelling of reaction kinetics and emission prediction (for example in terms of NO_x formation inside the cylinder, [16]).

The new developed surrogate mechanism is, of course, a simplification of reality. The surrogate mechanism can display the CO and H_2 oxidation behavior inside the exhaust manifold quite well, but it is still a simplified approach. With different boundary conditions and different time spans, the surrogate mechanism results are not necessarily realistic. Nevertheless, the mechanism is suitable for this application.

The oxidation-limit number N is a viable tool to investigate the oxidation limits of the exhaust gas. It can determine whether an increase in reactivity or mixing would lead to an increase in post-oxidation inside the exhaust manifold. In a future investigation, the usage of this indicator might lead to an additional gain in knowledge.

For further investigations, a 1D engine model with implemented 1D post-oxidation model can be used to evaluate the engine behavior due to post-oxidation in a wide range of engine operating points. This enables the possibility to rate the torque increasing effect of scavenging with a more profound approach than the discussed workarounds like e.g. an artificial increase of the turbines efficiency. In [1] a pre-study including map-based longitudinal dynamics RDE simulation is carried out. Since the used engine model is only well-tuned in the scavenging engine operating points and not in the entire engine map, the evaluation is made with respect to the amount and duration of scavenging moments instead of the influence of the 1D post-oxidation model on emissions and the overall fuel consumption due to a possible shift of the low-end torque (that enables the use of a different gear ratio). The pre-study shows that a bulk of the scavenging moments have a duration of less than 5 s. But there are also scavenging moments that last longer than 10 s. The longer the scavenging moment appears, the longer is the duration of increased heat release due to post-oxidation effects and also the time span of approaching not oxidized O_2 at the TWC. The approaching O_2 as well as the increased temperature of the exhaust gas might damage the TWC. Therefore, the post-oxidation model could be used to develop a gear shift strategy, that affects takes also the duration of scavenging moments into account to spare the TWC.

By using a full transient engine model, the heat release due to post-oxidation effects can be investigated with regard to the thermomanagement of the TWC. Since the TWC requires a certain temperature to work in a reliable way, a well thermomanagement is required to lower the emissions during e.g. a RDE driving cycle. The precise use of the post-oxidation effect could be key for such a thermomanagement. The 1D post-oxidation model is then a key element for a possible engine operating strategy development.

Bibliography

[1] *Post-Oxidation.* FVV Final Report, Project Number 1336, 2021.

[2] *GT-POWER.* Gamma Technologies, Version 2019.

[3] B. An, H. Suzuki, M. Ebisu, and H. Tanaka. Development of two-stage turbocharger system with electric supercharger. In *Proceedings of the FISITA 2012 World Automotive Congress*, volume 189 of *Lecture Notes in Electrical Engineering*, pages 147–155. Springer Berlin Heidelberg, Berlin, Heidelberg, 2013.

[4] L. Cai and H. Pitsch. Optimized chemical mechanism for combustion of gasoline surrogate fuels. *Combustion and Flame*, 162(5):1623–1637, 2015.

[5] M. Chiodi. *An innovative 3D-CFD-approach towards virtual development of internal combustion engines: Zugl.: Stuttgart, Univ., Diss., 2010.* Vieweg+Teubner Verlag / Springer Fachmedien Wiesbaden GmbH Wiesbaden, Wiesbaden, 2011.

[6] P. Dagaut, F. Lecomte, J. Mieritz, and P. Glarborg. Experimental and kinetic modeling study of the effect of no and so2 on the oxidation of co/h2 mixtures. *International Journal of Chemical Kinetics*, 35(11):564–575, 2003.

[7] R. Fairbrother, T. Leifert, R. Gande, and G. Salentinig. Accurate gas exchange and combustion analysis directly at the test bed. In *ASME 2012 Internal Combustion Engine Division Fall Technical Conference*, pages 667–676. American Society of Mechanical Engineers, 09232012.

[8] A. Fandakov. *A Phenomenological Knock Model for the Development of Future Engine Concepts.* Springer Fachmedien Wiesbaden, Wiesbaden, 2019.

J. Przewlocki, *Simulative Investigation of Post-Oxidation in the Exhaust Manifold of SI Engines*, Wissenschaftliche Reihe Fahrzeugtechnik Universität Stuttgart, https://doi.org/10.1007/978-3-658-36378-9

[9] T. Faravelli. Kinetic modeling of the interactions between no and hydrocarbons in the oxidation of hydrocarbons at low temperatures. *Combustion and Flame*, 132(1-2):188–207, 2003.

[10] A. J. Feneley, A. Pesiridis, and A. M. Andwari. Variable geometry turbocharger technologies for exhaust energy recovery and boosting–a review. *Renewable and Sustainable Energy Reviews*, 71:959–975, 2017.

[11] M. Fleiss, R. Burenius, G. Almkvist, and J. Björkholtz. The pneumatic turbocharger support system powerpulse. *MTZ worldwide*, 77(6):10–15, 2016.

[12] A. Frassoldati, T. Faravelli, and E. Ranzi. The ignition, combustion and flame structure of carbon monoxide/hydrogen mixtures. note 1: Detailed kinetic modeling of syngas combustion also in presence of nitrogen compounds. *International Journal of Hydrogen Energy*, 32(15):3471–3485, 2007.

[13] D. G. Goodwin, R. L. Speth, H. K. Moffat, and B. W. Weber. Cantera: An object-oriented software toolkit for chemical kinetics, thermodynamics, and transport processes, 2018.

[14] M. Grill. *Objektorientierte Prozessrechnung von Verbrennungsmotoren*. Universität Stuttgart, 2006.

[15] M. Grill, T. Billinger, and M. Bargende. Quasi-dimensional modeling of spark ignition engine combustion with variable valve train. SAE International, 2005.

[16] B. Kaal. *Phänomenologische Modellierung der stationären und transienten Stickoxidemissionen am Dieselmotor*. Springer Fachmedien Wiesbaden, Wiesbaden, 2017.

[17] B. Kaal, M. Grill, and M. Bargende. Transient simulation of nitrogen oxide emissions on diesel engines. In W. Siebenpfeiffer, editor, *Heavy-Duty-, On- und Off-Highway-Motoren 2015*, pages 174–188, Wiesbaden, 2018. Springer Fachmedien Wiesbaden.

[18] R. J. Kee, F. M. Rupley, and J. A. Miller. The chemkin thermodynamic data base.

[19] A. Kéromnès, W. K. Metcalfe, K. A. Heufer, N. Donohoe, A. K. Das, C.-J. Sung, J. Herzler, C. Naumann, P. Griebel, O. Mathieu, M. C. Krejci, E. L. Petersen, W. J. Pitz, and H. J. Curran. An experimental and detailed chemical kinetic modeling study of hydrogen and syngas mixture oxidation at elevated pressures. *Combustion and Flame*, 160(6):995–1011, 2013.

[20] C. Koehlen, E. Holder, and G. Vent. Investigation of post oxidation and its dependency on engine combustion and exhaust manifold design. In *SAE Technical Paper Series*, SAE Technical Paper Series. SAE International400 Commonwealth Drive, Warrendale, PA, United States, 2002.

[21] S. Lecheler. *Numerische Strömungsberechnung*. Springer Fachmedien Wiesbaden, Wiesbaden, 2014.

[22] T. Lu and C. K. Law. A directed relation graph method for mechanism reduction. *Proceedings of the Combustion Institute*, 30(1):1333–1341, 2005.

[23] Mehl M., W.J. Pitz, C.K. Westbrook, H.J. Curran. Kinetic modeling of gasoline surrogate components and mixtures under engine conditions. In *Proceedings of the Combustion Institute 33*. 2011.

[24] C.-D. Munz and T. Westermann. *Numerische Behandlung gewöhnlicher und partieller Differenzialgleichungen: Ein interaktives Lehrbuch für Ingenieure*. Springer Berlin Heidelberg, Berlin, Heidelberg, 3. aufl. 2012 edition, 2012.

[25] K. E. Niemeyer, C.-J. Sung, and M. P. Raju. Skeletal mechanism generation for surrogate fuels using directed relation graph with error propagation and sensitivity analysis. *Combustion and Flame*, 157(9):1760–1770, 2010.

[26] N. Peters. *Turbulent combustion*. Cambridge monographs on mechanics. Cambridge University Press, Cambridge, 2000.

[27] E. Ranzi, A. Frassoldati, R. Grana, A. Cuoci, T. Faravelli, A. P. Kelley, and C. K. Law. Hierarchical and comparative kinetic modeling of laminar

flame speeds of hydrocarbon and oxygenated fuels. *Progress in Energy and Combustion Science*, 38(4):468–501, 2012.

[28] A. Schmid, M. Grill, H.-J. Berner, and M. Bargende. Transient simulation with scavenging in the turbo spark-ignition engine. *MTZ worldwide*, 71(11):10–15, 2010.

[29] D. Serrano, J. Lavy, A. Kleemann, S. Zinola, J.-P. Dumas, S. Le Mirronet, and D. Heitz. Post oxidation study during secondary exhaust air injection for fast catalyst light-off. In *SAE Technical Paper Series*, SAE Technical Paper Series. SAE International400 Commonwealth Drive, Warrendale, PA, United States, 2009.

[30] P. Skarke. *Simulationsgestützter Funktionsentwicklungsprozess zur Regelung der homogenisierten Dieselverbrennung*. PhD thesis, Wiesbaden, 2017.

[31] A. Stagni, A. Frassoldati, A. Cuoci, T. Faravelli, and E. Ranzi. Skeletal mechanism reduction through species-targeted sensitivity analysis. *Combustion and Flame*, 163:382–393, 2016.

[32] Thorsten Günther, M. Grill, and M. Bargende. A simulative study for post oxidation during scavenging on turbo charged si engines. In *WCX World Congress Experience*. SAE, 2018.

[33] R. Tromellini, J. Przewlocki, F. Cupo, M. Bargende, and M. Chiodi. Analysis of scavenging air post-oxidation by means of 3d-cfd simulation including reaction mechanism. In M. Bargende, H.-C. Reuss, and A. Wagner, editors, *20. Internationales Stuttgarter Symposium*, Proceedings, pages 421–437. Springer Fachmedien Wiesbaden, Wiesbaden, 2020.

[34] S. R. Turns. *An introduction to combustion: Concepts and applications*. McGraw-Hill, Boston, 3. ed. edition, 2011.

[35] J. Warnatz, R. W. Dibble, and U. Maas. *Combustion: Physical and Chemical Fundamentals, Modeling and Simulation, Experiments, Pollutant Formation*. Springer-Verlag Berlin Heidelberg, Berlin, Heidelberg, 4th edition edition, 2006.

[36] B. Weigand, J. Köhler, and J. von Wolfersdorf, editors. *Thermodynamik kompakt*. Springer Berlin Heidelberg, Berlin, Heidelberg, 2016.

The manufacturer's authorised representative in the EU is Springer Nature Customer Service Centre GmbH, Europaplatz 3, 69115 Heidelberg, Germany. If you have any concerns regarding our products, please contact ProductSafety@springernature.com

Printed and bound by CPI Group (UK) Ltd, Croydon, CR0 4YY
24/07/2026
02175971-0002